Carl Hering

Universal Wiring Computer

Carl Hering

Universal Wiring Computer

ISBN/EAN: 9783744675277

Printed in Europe, USA, Canada, Australia, Japan

Cover: Foto ©berggeist007 / pixelio.de

More available books at **www.hansebooks.com**

UNIVERSAL

WIRING COMPUTER

FOR DETERMINING

The Size of Wires for Incandescent Electric Lamp Leads

AND FOR DISTRIBUTION IN GENERAL,

Without Calculation, Formulæ or Knowledge of Mathematics

WITH SOME NOTES ON WIRING AND A SET OF

AUXILIARY TABLES.

BY

CARL HERING

*

NEW YORK:

THE W. J. JOHNSTON COMPANY, Ltd.

1891

PREFACE.

In submitting to the public the accompanying new system for determining the size of leads without calculations, the author desires to say that he has endeavored to make the charts as simple and practical as possible; but, as in other new departures, it is possible that such proportions as the dimensions of the scales, their ranges, the size of the charts, the limit of accuracy, etc., might be more advantageously chosen so as to bring the average values to the best parts of the charts. As the values for the usual determinations have such very wide limits, it is difficult to determine on the best proportions of the charts except by long and repeated use in practice under widely differing circumstances. Since this can best be done by the aid of those using this system under different circumstances, the author appeals to those using these charts to aid him in finding the most convenient proportions, by suggesting to him any changes in the present proportions gained from actual experience with the charts. Such changes will, if practicable, be embodied and duly acknowledged in subsequent editions, of which copies will be sent to those to whose kindness the author owes the changes.

As its title indicates, this book is intended to facilitate the computing of the size and quantity of wire used for wiring; it is not a treatise on wiring, but assumes a knowledge of wiring on the part of the reader. It is intended for a book of reference, and not for a book of instruction. The auxiliary tables, which were almost all calculated for this book, are limited to those which wiremen frequently have occasion to refer to.

. The author is indebted to his friend Richard W. Davids for some practical suggestions, and to the ELECTRICAL ENGINEER for the use of some of the illustrations.

CARL HERING.

Philadelphia, April, 1891.

CONTENTS.

UNIVERSAL WIRING COMPUTER.

INTRODUCTION.

THE determination of the proper size of the wire for distributing current for incandescent lighting, is burdened with the use of formulæ having "constants" varying with each style of lamp; these constants mean different things, depending on which formula is used; furthermore many wiremen and contractors may not know how this constant is determined, and therefore they cannot deduce it themselves if they have forgotten it, or if they have to wire for a different make of lamp. Such formulæ and constants are therefore often unsatisfactory for all cases except for daily work with one particular make of lamp. Even then there is no small amount of calculation necessary to make a proper determination of the wiring of a building; the natural consequence is that much of the wiring is a mere guess as to the size of wire, and it is a matter of chance whether this guess is a good one or a bad one. The sizes of wire may be so widely different for differing conditions, that a "guess" is more likely to be a bad one, except, perhaps, in the unfrequent case of a person making very many determinations daily for the same make of lamp; even in such cases it is well to check the results by a proper determination. The competition among contractors for wiring is getting to be so great that it will be the one who makes the most economical determination of the sizes of wire, who will be able to outbid his competitors who may either waste wire in making it too large, or have to add an additional wire afterwards in case it was too small.

The cause of much of this "guessing" is doubtless due to the fact that it requires no small amount of figuring to make even an approximate determination of the sizes of the wire. It is to diminish this work that the author has devised the accompanying charts, by means of which all such determinations are made at a glance, more readily even than if the values were looked up in tables (if such tables existed), which would necessarily have to be very bulky and cumbersome, in order to cover such a wide range as that required for the general practice.

EXPLANATION OF THE CHARTS.

GENERAL.—These charts will give directly and without calculation or the use of formulæ, the gauge number or cross-section in circular mils of leads for any number of lamps of any make, at any distance and for any loss. There are three charts with different scales, covering the following ranges:

Few lamps at short distances.
Few lamps at long distances.
Many lamps at short distances.

Also a blank chart which can be filled out for any special ranges, as will be described below. The ranges overlap somewhat, so that if the values sought for are on awkward parts of the charts, they will probably be found in better parts on one of the other charts. They cover the ranges for house wiring, for large or small houses, and give the results with a degree of accuracy which is far greater than is necessary on account of the wide limits between the standard sizes of wire in the market; a greater accuracy than this would be absurd, as one cannot generally obtain the wire for any but the regular sizes, and not even for all of these. For many lamps at a great distance, a small error would make a great difference in the cost of the wire. For such cases the wire must be calculated by means of the usual formula, for which see page 19.

HOW TO USE THE CHARTS.—The vertical scale just below the center represents the current in amperes for one lamp. Find the current of the particular make of lamp on this scale, and follow it horizontally to the left until it intersects the diagonal representing the desired loss in volts (see broken line on charts); from this intersection follow the corresponding vertical line until it intersects the diagonal in the upper left hand portion, representing the desired number of lamps; from this intersection follow horizontally to the right to the next set of diagonals representing the distances in feet (not the length and return, but merely the distance one way), and from this intersection follow down vertically to the scale which gives the circular mils, as also the B. & S. (American)

(3)

wire gauge numbers. An example is worked out on each chart and indicated on the chart by a broken line.

It should be noticed that the loss is given in *volts*, and not in per cent., except for a 100-volt lamp, for which the loss in per cent. and in volts is the same thing. For any other voltage, if the loss is given in per cent., find the number of volts which this represents before starting to use the chart. This is done by multiplying the voltage of the lamp, say 75 volts, by the per cent., say 2 per cent., and divide by 100; thus, $75 \times 2 \div 100 = 1.5$ volts.

HINTS AND MODIFICATIONS.

FOR ONE PARTICULAR MAKE OF LAMP.—If, as is generally the case, a large number of determinations are to be made for one particular make of lamp, the work can be shortened considerably by laying off with care, on the first scale, the current for that lamp, and then with a lead pencil or red ink draw a bold horizontal line across to the left. The intersections of this line with the volt diagonals will then be the starting points for the different losses. The numbers which the diagonals represent can then be transferred to this line for convenience.

FOR ONE PARTICULAR LOSS.—If, besides using the same lamp the loss is also the same for a large number of determinations, which is very often the case, then draw a second red line, or bold pencil line, vertically upward across the "lamp" diagonals, then these intersections (in the upper left hand field) will be the starting points for all determinations, thus simplifying the work by reducing it to one-half. It is recommended in this case to transfer the numbers representing the lamps to the intersections of this new line, with the respective diagonals in that field, as these intersections form the starting points.

STANDARD SIZES OF WIRES.—The work is still further simplified by the vertical dotted lines in the right hand field which have been drawn through the gauge numbers on the scale which represent the standard B. &. S. sizes of wire. This facilitates following the vertical lines down to the scale, thus reducing the amount of work still more.

LOSS IN PER CENT. INSTEAD OF VOLTS.—If it is preferable to have the losses read in per cent. instead of in volts, the change can be made by calculating what percentages are represented by each of the volt lines, and marking them accordingly. But such figures will be correct only for lamps of that same voltage, and for no other.

INTERPOLATING.—For values lying between two diagonals, or when new diagonals are drawn for special values (as, for instance, for one standard loss in volts), notice that in the lower left hand field the distances between the diagonals should be measured on a vertical scale on which they are proportional to the volts; for instance, a diagonal representing $1\frac{1}{4}$ volt would be half way between that for 1 volt and that for $1\frac{1}{2}$ volt, measured on any *vertical* line, and not on a horizontal line nor on the arc of a circle. The same thing is true of the upper left hand field (lamps), namely, that the vertical scale is quite regular. In the upper right hand field (feet), it is the horizontal scale and not the vertical which is regular.

CHANGING THE SCALES.—The following points are worth remembering. The number of lamps and the distances in feet are interchangeable. It may be more convenient sometimes to use lamps for feet and feet for lamps; both give the same result. Furthermore, either of these two may be divided or multiplied by 2, or 10, or 100, etc., if the other one is correspondingly multiplied or divided by the same factor. For instance, 1 lamp at 400 feet is the same as 2 lamps at 200 feet, or 4 lamps at 100 feet. Sometimes one or the other of these alternatives is more convenient to find. With the volt scale, however, it is different; if the volt figures are multiplied by two, for instance, the lamp figures (or the feet) must be multiplied (not divided) by two also; for instance, for a 1-amp. lamp and a $\frac{1}{2}$-volt loss, the intersection falls off the chart; but by using the 1-volt diagonal instead, and doubling the number of lamps (or the feet), the final result will be the same. Such changes are rarely necessary, on account of the different ranges of the different charts; but it may often be less trouble to take such an alternative than to turn to another chart.

SPARE CHARTS.—A spare chart has been added on which the lines are identical with those on the other charts. This may be filled out with figures so as to cover any special work, as, for instance, for the three-wire system, for motor work, or perhaps for improvements on the ranges of the scales of the other charts.*

* See Preface.

The two preceding paragraphs will explain in what proportions the numbers may be changed without changing the lines themselves. Spare charts may be obtained from the publisher.

LAMPS OF DIFFERENT CANDLE-POWERS.—If lamps of different candle-power (that is, having different currents) are mixed and are on the same circuit, they must either all be reduced to their equivalent in terms of the same lamp, or else if there are only two or three kinds, the leads may be determined in circular mils (not in gauge numbers) for each batch of like lamps separately, and the sum of all the circular mils taken, from which sum the gauge numbers are then found from a table or from the double scale on the chart.

POWER LEADS.—For the distribution of power, start with the line (near the bottom of the chart) representing a *one-ampere* lamp, then the numbers representing lamps in the upper left hand field will represent amperes of current. The current in amperes corresponding to the horse-power must, of course, be determined first from the horse-power and the volts (see table of horse-power equivalents, pages 36 and 37).

THREE-WIRE SYSTEM.— If the wiring is to be done for the three-wire system in which three wires of like size are used in place of two, the cross-section of each will be one-fourth as great as that for the ordinary system. Instead of finding the cross-section from the charts and dividing it by four, and then finding the gauge number from a table, it is much simpler to proceed as before, but taking either one-fourth the number of lamps or one-fourth of the distance, or four times the loss. By doing it in this way the size of wire is obtained directly without the use of any table, while the only calculation necessary is merely a mental one.

OTHER USES OF THE CHARTS.—The charts may be used backward, so to speak, by starting with a given size of wire and working backward to find what the loss will be for a given number of lamps at a given distance. In the same way, the allowable number of lamps or the distance may be determined if the other quantities are given. In general, any one of the quantities may be found if all the others are given; the general rule in that case is to start from the beginning and end of the chart simultaneously, and continue as usual until the two lines which one is following intersect in the common field which contains the diagonals representing the quantity looked for; that diagonal which passes through or nearest to this intersection represents the number sought for. For instance, how many .775-ampere lamps will a No. 11 wire carry, to a distance of 50 feet, with a loss of 1 volt? See the first chart, broken line. Starting with the line representing a .775-ampere lamp, follow it to the 1 volt loss line; thence up into the field representing lamps; then begin with the intersection of the dotted line representing a No. 11 wire and the 50 feet line, and follow backward (see broken line) to the lamp field; where it crosses the other line, find what diagonal passes through this point; this diagonal, namely 10 lamps, is the required number of lamps.

AUXILIARY TABLES.—At the end of the book there are some tables which will frequently be found useful in connection with wiring determinations.

MANY LAMPS AT A GREAT DISTANCE.—If the leads are to be determined for many lamps at a great distance, a small error in the determination signifies a considerable difference in cost of the wire; the computation must therefore be made more accurately. To do this would require a chart of very great size. It is therefore preferable to calculate such exceptional determinations by means of one of the following rules:

If the total current is given: *multiply the total current by the distance in feet and by 21.21, and divide by the loss in volts;* the result will be the required cross-section of the leads in circular mils.

If the current per lamp is given: *multiply the current per lamp by 21.21; this gives the " constant"; multiply the number of lamps by the distance in feet and by this " constant," and divide by the loss in volts;* the result will be the required cross section in circular mils.

The gauge numbers corresponding to these cross sections will be found in the tables at the end of the book, page 23. For very large cross sections a·table is given showing what sizes of wires bunched together will make this cross-section. (See page 28).

BASIS OF THE CHARTS.—The basis of these charts (as also that of the tables and formulæ in this book) is a resistance of 10.61 legal ohms per mil foot of copper wire. In terms of the Matthiessen standard suggested by the Committee of the American Institute of Electrical Engineers (namely, 9.612 legal ohms per mil foot at 0° C.), this is equivalent to the resistance at about 75° to 80° F. As pure copper of the present time sometimes has even less resistance than that referred to in this standard, it is thought

that the value chosen for these charts and tables represents a fair
.value for the resistance of good copper at the average normal temperature.   As the accidental differences in the actual diameters of
the wires introduce errors far greater than a slight difference in
the assumed standard conductivity, it would not be reasonable to
attach much importance here to great precision in the assumption
of the standard.   All that is necessary here is to select the fairest
possible value for actual practice, to state what this value is, and
to have it the same throughout this whole set of charts, tables and
formulæ.

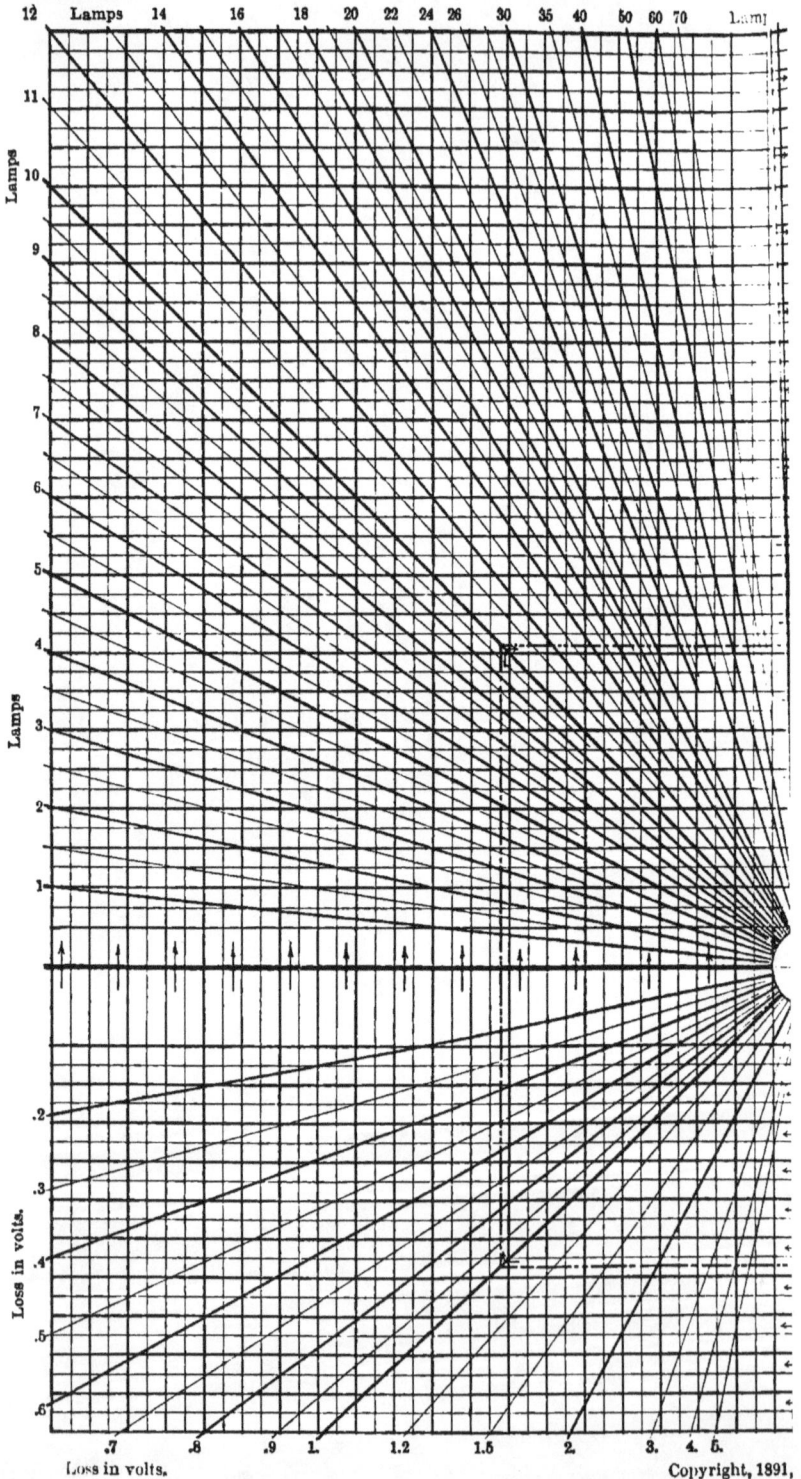

Copyright, 1891,

3
4
5
6
7
3
)
)

Ampere. Single Lamp Current.

FEW LAMPS AT SHORT DISTANCES.

Rule for using the chart:

Follow the general direction of the broken line and the arrows, from one set of diagonals to the next.

EXAMPLE: What size wire is required for 10 lamps of .775 amperes each, at 50 feet, for a loss of 1 volt?

SOLUTION : Starting with the current for 1 lamp, .775 amperes (see scale below center), follow it (see broken line and arrows) to the left, until it intersects the diagonal representing 1 volt loss ; thence up to the diagonal representing 10 lamps ; thence to the right to the diagonal representing 50 feet, and from here down to the scale of the circular mils or gauge numbers, on which the reading is found to be about 8,200 circular mils, or a No. 11 B. & S. wire.

For a more detailed explanation, abbreviated methods and general hints, see text.

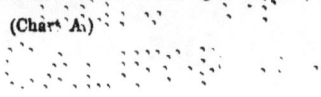

(Chart A)

ARL HERING.

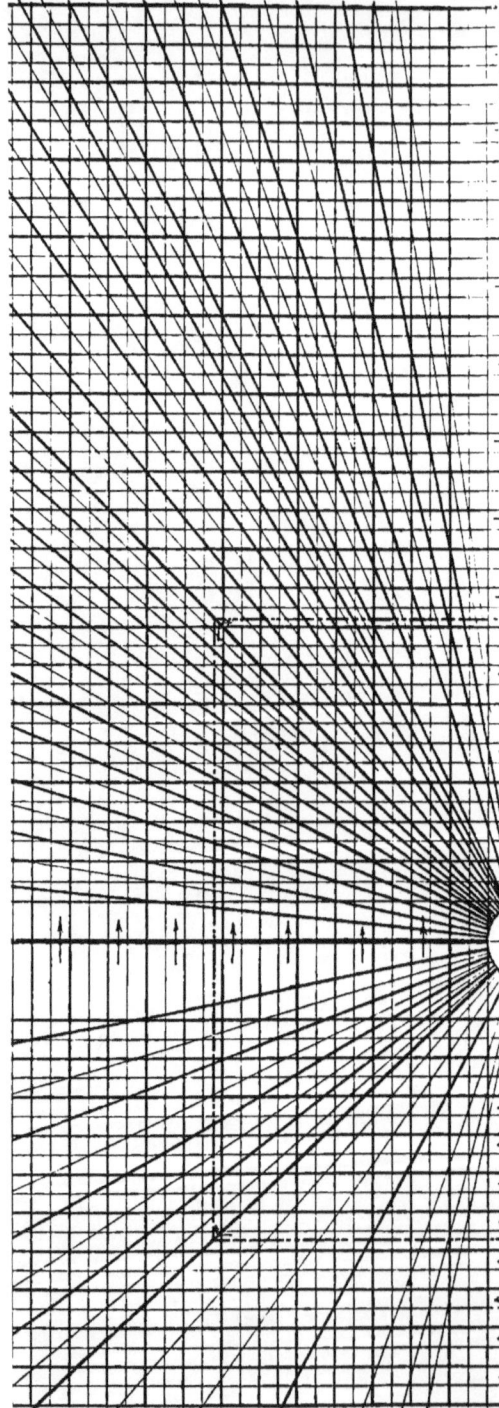

18 20 22 24 26 30 35 40 50 60 70 Lamps

2.   2.4   3.   4.   6. 8. 10.

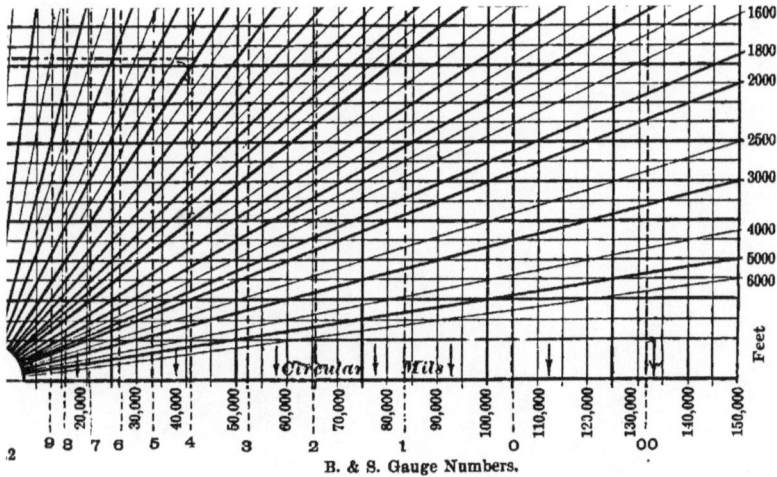

Right-side axis labels (Feet): 1600, 1800, 2000, 2500, 3000, 4000, 5000, 6000

Circular Mils (along the diagonal scale): 20,000 · 30,000 · 40,000 · 50,000 · 60,000 · 70,000 · 80,000 · 90,000 · 100,000 · 110,000 · 120,000 · 130,000 · 140,000 · 150,000

B. & S. Gauge Numbers: .2 · 9 8 7 6 5 4 · 3 · 2 · 1 · O · OO

Left-side axis label (Ampere. Single Lamp Current.): .3 .4 .5 .6 .7 .8 .9 .0 .1 .2

FEW LAMPS AT LONG DISTANCES.

Rule for using the chart:

Follow the general direction of the broken line and the arrows, from one set of diagonals to the next.

EXAMPLE: What size wire is required for 10 lamps of .775 amperes each, at 500 feet, for a loss of 2 volts?

SOLUTION : Starting with the current for 1 lamp, .775 amperes (see scale below center), follow it (see broken line and arrows) to the left, until it intersects the diagonal representing 2 volts loss; thence up to the diagonal representing 10 lamps; thence to the right to the diagonal representing 500 feet, and from here down to the scale of the circular mils or gauge numbers, on which the reading is found to be about 42,000 circular mils, or a No. 4 B. & S. wire.

For a more detailed explanation, abbreviated methods and general hints, see text.

(Chart B.)

ARL HERING.

1.4   1.6   1.8   2.   2.4

Loss in volts.

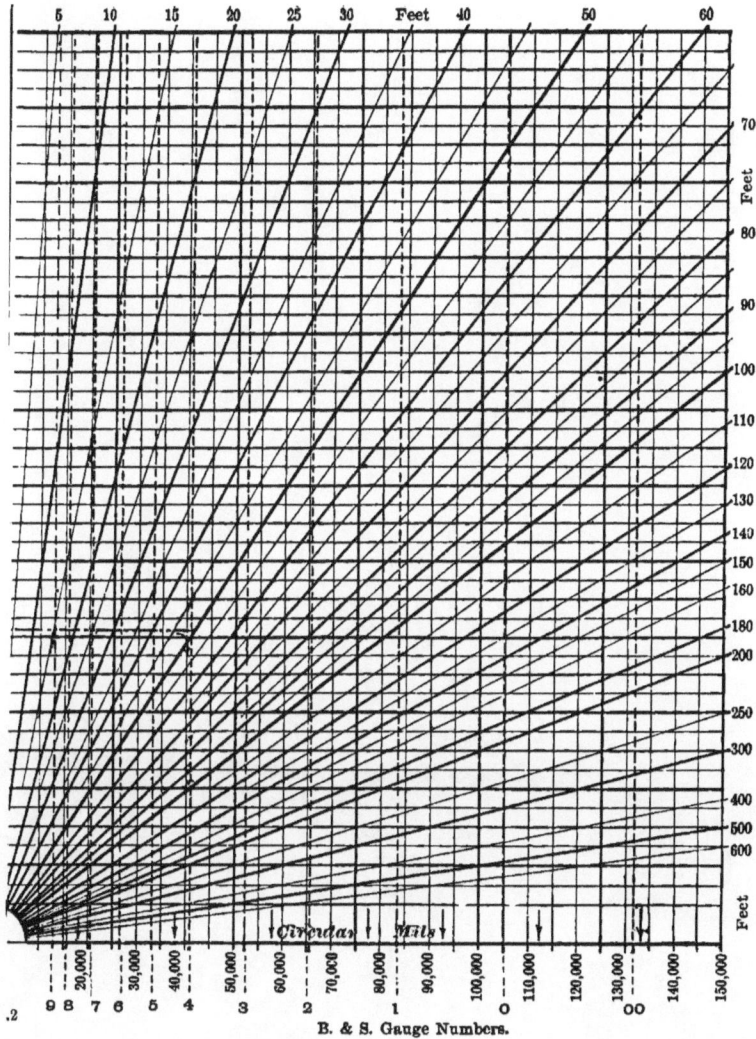

Top axis: 5 10 15 20 25 30 Feet 40 50 60

Right axis (Feet): 70 80 90 100 110 120 130 140 150 160 180 200 250 300 400 500 600

Bottom scale (Circular Mils): 20,000 30,000 40,000 50,000 60,000 70,000 80,000 90,000 100,000 110,000 120,000 130,000 140,000 150,000

B. & S. Gauge Numbers: 9 8 7 6 5 4 3 2 1 0 00

Left axis (Ampere. Single Lamp Current): .2 .3 .4 .5 .6 .7 .8 .9 .0 .1 .2

MANY LAMPS AT SHORT DISTANCES.

Rule for using the chart:

Follow the general direction of the broken line and the arrows, from one set of diagonals to the next.

EXAMPLE: What size wire is required for 100 lamps of .775 amperes each, at 50 feet, for a loss of 2 volts?

SOLUTION : Starting with the current for 1 lamp, .775 amperes (see scale below center), follow it (see broken line and arrows) to the left, until it intersects the diagonal representing 2 volts loss; thence up to the diagonal representing 100 lamps; thence to the right to the diagonal representing 50 feet, and from here down to the scale of the circular mils or gauge numbers, on which the reading is found to be about 42,000 circular mils, or a No. 4, B. & S. wire.

For a more detailed explanation, abbreviated methods and general hints, see text.

(Chart C.)

ARL HERING.

(Chart D.)

CALIFORNIA

DISTRIBUTION OF INCANDESCENT LAMP LEADS.

IN the ideal system of wiring for incandescent lamps (or motors) in multiple arc, there are two requirements, assuming that the potential is kept constant at the source: *First*, that each lamp should have the same potential at its socket as all the other lamps, when all are burning at once; *secondly*, that this potential should remain constant at each lamp, when the other lamps are turned off. In some cases, as for factories, street lights, store lights, etc., only the first requirement need be complied with; in other cases, as in dwellings, theaters, etc., both conditions must be provided for. The second condition is the one most difficult to provide for, and it necessarily includes the first. A number of systems of running the leads will comply with the first condition, but to meet the second condition there is only one ideal system.

In general, it would be quite impracticable to comply strictly with either of these conditions, and therefore a slight margin of variation at the different lamps is usually allowed; the amount of such allowable variation, being necessarily different for different conditions, must be chosen by the judgment of the engineer. This variation is due to the different losses in potential in the leads to the different lamps. This variation is, therefore, not identical with the loss in the leads, but it is the differences between these losses when the losses are not precisely the same to all lamps.

The amount of wire used increases very rapidly as the allowable loss is diminished; for a 1 per cent. loss, for instance, the amount of wire by weight is double what it would be for a 2 per cent. loss. On the other hand, if the lamps must be capable of being turned off one by one, the life of the lamps in the general systems will diminish rapidly as this allowable loss is increased, because the unavoidable differences between the losses to different lamps, that is, the variation, increases. It is, therefore, a choice between two evils. As the conditions are quite different in the

(9)

case when the lamps may be turned off one by one, than when
they always burn together, the two cases must be considered separately and should not be confounded with each other; the former,
of course, includes the latter, and is, therefore, merely an additional condition to the latter.

The case in which all the lamps are either burning or turned
off together, is by far the simpler of the two.   In the simple case
shown in Fig. 1, the difference between the potential at the nearest

FIG. 1.

lamp and that at the farthest, is merely the amount lost in the
length of wire between the two lamps, and it is entirely independent of the amount lost between the dynamo (or center of distribution) and the first lamp; this latter loss may, therefore, be made
as great as desired, as far as the lamps are concerned.   In this
case, therefore, the loss from the dynamo to the first lamp may be
made anything that is desired, but the wire between the first and
last lamp must be so large that the loss on that portion does not
exceed the allowable variation for the lamps, say about 1 per cent.;
or at most, 2 per cent.   If this portion of the circuit is so long
that it would require a very large wire, then the lamps are often
divided into two or more groups, as shown in Fig. 2, each group

FIG. 2.

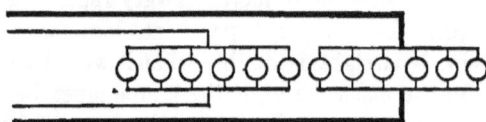

being supplied or fed by a separate set of leads; these two separate sets of leads must then be calculated for the same loss as
before between the dynamo and the lamps, thus requiring the
longer ones to be much thicker than the shorter ones, as shown.
The choice between the dispositions shown in Figs. 1 and 2 depends entirely on whether the wire between the first and the last
lamp must be so thick, owing to the allowable difference between
the lamps, that it would be cheaper to divide the leads to dynamo
into two parts; this can be determined only by calculating both
cases.   In Fig. 1 it might, under special circumstances, be quite

rational to connect the lamps by a much thicker wire than that leading to the dynamo, even though the latter carries a greater current. The disposition shown in Fig. 2, of running separate sets of leads to different groups of lamps, applies equally well to groups of lamps in different directions from the center of distribution, and in this sense it is one of the most frequent and best systems of distribution.

Another system, but applicable only in special cases, is that shown in Fig. 3, in which the two leads from the dynamo divide,

FIG. 3.

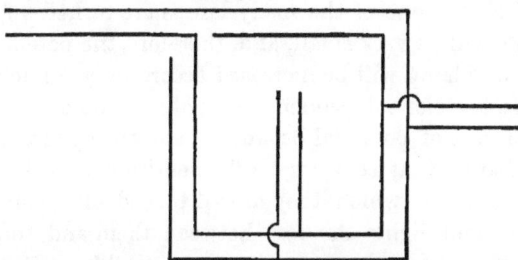

one going in one direction around a rectangle, and the other going in the reverse direction. No matter what the loss or the size of the wire, all the lamps between this pair of leads will have the same potential, provided the positive and negative leads are both of the same size, and provided all the lamps are turned on and off together; if the lamps are turned off one by one, the potential will no longer be constant.

To recapitulate, it will be seen that when the lamps of a group are all turned on or off together, and not individually, the distribution is simple, requiring only that the *difference* between the potential at the nearest and the farthest lamp on the same leads shall not exceed the allowable variation of 1, 2 or even 3 per cent. (in which case the lamps are entirely independent of the loss between them and the dynamo or center of distribution), and that if there are a number of such groups connected to the same dynamo or center, the loss from dynamo to lamp must be the same for each group. In the latter case the groups will be entirely independent of each other, and may be turned off or on as individual groups, provided their leads do not join those of any other group on their way to the dynamo. In other words, groups having independent connections to the dynamo are independent of each other, and

may be turned off or on as groups. It is assumed, of course, that
the dynamo is self-regulating.

Taking up the other case, in which the individual lamps are to
be turned off and on, the problem is quite a different one. Referring again to Fig. 1, the loss of potential from each lamp to the
dynamo or source, depends on the total current in the common
leads and on the resistance of these leads; these losses, therefore,
remain constant only as long as the total current is constant; if
one lamp is turned off, the total current becomes less, and, therefore, the loss to each remaining lamp becomes less, and *vice versa*.
Finally, if all but one of the many lamps are turned off, the loss
in the leads will be very small, and, therefore, the potential at the
last remaining lamp will be increased accordingly. Each individual lamp is, therefore, dependent not only on the others, but also
on the total loss of potential between it and the dynamo. It is in
the latter feature that this case differs entirely from the first case
described above, in which they are all turned off or on together.
For independent lights the loss between them and the dynamo
must, therefore, be made as small as practicable, as it affects the
steadiness and life of the lamps. For this reason it requires, in
general, relatively thicker wire for independent lamps than for
groups, provided the distance to the dynamo is sufficiently great
to make a difference.

Suppose, in Fig. 1, there are 100 lamps and the loss from the
dynamo to the first lamp is four volts when all are burning; then
if all but one are turned off, the voltage of that one will be about
four volts in excess of what it should be. In order to save the
lamps from part of this strain, the voltage of the dynamo may be
so chosen that when all are burning they will be two volts below
the normal, and when only one is burning it will be two volts
above, the difference remaining, as before, four volts. If, as in a
dwelling, the full number of lights burning is the exception, and
a few lights the rule, then the potential at the dynamos may be
chosen so that it is the proper amount at the lamps when the average number is turned on.

As the potential at the lamp depends on the total current in
the leads, it follows that the ideally perfect system of independent
lamps is to have a separate pair of leads for each lamp back to
the dynamo (or center of distribution). Each pair of leads is then
calculated so as to have the required loss for its lamp. Such an

ideal system is, however, not practicable, as a rule, but the general
rule may be laid down, that the nearer a system approaches to
this ideal, the better it is.   For instance, comparing Figs. 1 and 2,
in each of which there are twelve lamps, the second approaches
more nearly to the perfect system, and the lamps in Fig. 2 are
subjected to only *half* as great a variation in potential as those in
Fig. 1, when all but one are turned off.   It follows from this rule
that the distribution is better, the more a circuit is branched, the
nearer such branch connections are to the dynamo, and the larger
the number of independent leads to the dynamo.   Such distribution is better, not only because the lamps are subjected to less
variation of potential, but for the same allowable variation of, say
2 per cent. at the lamp, the total loss from the dynamo to the
lamps may be chosen much greater than in other systems and
thereby saving wire, for it is evident that in the ideal system the
loss from the dynamo to the lamps may be made anything desired, without making the lamps dependent on each other ; their
dependence on each other varies with and is proportioned to, the
number of lamps on one wire, and the distance from the dynamo
to the junction of their individual wires.

It is sometimes thought that the ideal system may be carried
out by calculating the leads to each lamp or groups of lamps separately and then bunching all the wires running parallel into one
common wire having a cross-section equal to the sum of all the
smaller wires combined.   But this is an error and may result in a
worse distribution than if it had been calculated on the usual
plan.   As soon as two wires are metallically connected they become
one and the same wire from there to the dynamo.

To recapitulate : When the lamps are to be cut off independ-
ently they are dependent on each other and on the loss of potential
between them and the dynamo *in so far as they are connected to
common leads.*   The leads should therefore be split up as much as
practicable, and the total loss should be divided so as to have the
greater part in the small individual branch wires, and the smaller
part in the larger main wires.

To calculate the wires for a building with independent lamps,
lay them out so as to approach as much as practicable to the
best distribution as described above, making common mains as
short as possible, and individual branch wires as great a proportion of the whole as possible ; then determine on the total loss, for

instance, four volts, and divide it amongst the leads so as to have as small a part as practicable (say one volt) on the common main, and the other part (three volts) again divided, if necessary, on the distributing branch wires. Calculate the size of each wire from the number of lamps supplied by it, and from this portion of the total loss allowed for that part of the whole lead. The lamps will then be dependent on each other only in so far as they are on common wires, and to an amount that other lamps effect the loss only on this common wire.

To illustrate some of the points mentioned above by an actual (exaggerated) case, suppose the leads for four lamps, a, b, c, d, Fig. 4, be subdivided as shown, and suppose the total loss of 8 volts be

FIG. 4.

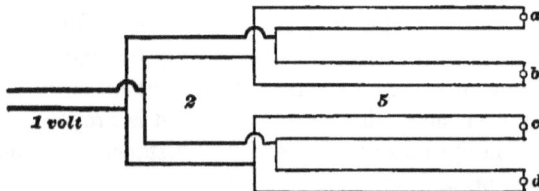

FIG. 5.

divided into 5, 2 and 1, as indicated, on the separate mains and branches; the relative distances being in the proportions of the diagram. The loss is proportionately small on the common mains and large on the individual branches. Now, taking any one lamp, as a, its voltage will be increased as follows: With b turned off, $1\frac{1}{4}$ volt; with c or d turned off, $\frac{1}{4}$ volt; with c and d both turned off, $\frac{1}{2}$ volt; with b, c and d turned off, $1\frac{3}{4}$ volt. This shows that lamp a is dependent on the others in proportion as it is on common mains with them, and on the loss of volts on the common mains, which is small in this case.

Now, for the sake of comparison, let the four lamps be supplied by a single pair of mains, as in Fig. 5, with the same loss of 8 volts. Turning off one lamp increases the voltage of the others 2 volts; with two lamps turned off, 4 volts; and with three lamps off, 6 volts. This shows how very great the difference is, namely, a maximum of 6 volts in Fig. 5, as compared to $1\frac{3}{4}$ volts in Fig. 4.

The weight of wire in Fig. 4 is only slightly higher, namely, as 23 to 20.   If now the wire in Fig. 5 be made larger, so as to have the same maximum *variation* in volts as in Fig. 4, namely, 1¾ volt, the total loss would have to be 2 volts, and this would increase the weight of wire to about three times that in Fig. 4, showing the advantage in subdividing the leads, aside from the fact that Fig. 4 is a distribution (as the lamps might just as well be at the same distance in different directions), while that in Fig. 5 is not. The actual figures will, of course, vary greatly under different circumstances, and no general statement can be made regarding the amount of gain.

In referring to the dynamo in the above deductions it was understood to mean the place from which distribution begins, that is, the center of distribution, or the common point at which the potential is kept constant.   In wiring large buildings or spaces it is usual to run a pair of large mains to a central point from which distribution begins; this pair of mains, provided it is the only one from the dynamo, is not included in the above discussion, as it is supposed that the dynamo is so regulated as to keep a constant potential at the far ends of this pair of mains, that is, at the center of distribution; if the dynamo does not do this, or if there is more than one pair of such mains, then it brings the center of distribution back to the dynamo, thus making these mains part of the distribution.

A great mistake is often made in supposing that a dynamo can keep the potential constant at more than one distant center of distribution, without special apparatus at the dynamo.   This refers, of course, to a system of independent lamps.   Suppose all lamps are turned on at one center, and only one lamp is on at the other, this lamp will be run too high, as the dynamo must be kept at the same high potential on account of the lamps on the other center. It can be accomplished only in one of two ways, first, approximately, by making the loss on the mains very small; secondly, by regulators in each of the original branches from the dynamo.

It has been suggested to put lower voltage lamps at the most distant centers, and higher voltage lamps at the nearer ones, on account of the greater loss in the longer mains.   It is a question whether this is practicable, for a number of reasons.   A lower volt lamp requires a greater current and, for this reason alone, a larger wire.   It is not a good practice to have lamps of differing voltages in stock for one and the same building or installation, unless there is a reliable person to take charge of their proper placing.

In the three-wire system there are practically two lamps in series, and, therefore, the current need be sent out and back only once for every two lamps; this requires but half the wire (in cross-section) otherwise necessary for the same number of lamps. Furthermore, the loss of volts is divided between two lamps, and it can therefore be made twice as great as in the simple system; this halves the quantity of wire again, making the total one-quarter as great as for the two-wire system. To carry the current for any lamp which may not at the time have another in series with it, a third or neutral wire is laid, which, in wiring buildings, is usually made the same size as the other two; this increases the wire by a half, making the total three-eighths of that required for the simple system. To calculate the leads for the three-wire system, proceed as in the simple system and divide the cross-section obtained by four, using three wires of this cross-section. The same result would, of course, be obtained by using one-quarter the number of lamps, or one-quarter the distance, or four times the loss.

FUSIBLE CUT-OUTS.

The general principle of safety or fusible cut-outs is that they protect from a dangerous excess of current those wires which are *beyond* them, as distinguished from the wires between them and the dynamo, which are not protected by the fuses. They should therefore always be placed at the *beginning* of a wire (that is, at the end toward the dynamo) and not at the lamp end. Furthermore, they should be made so small that they protect the *smallest* wire lying beyond them up to the next fuse; this is not infrequently overlooked, and may be a source of great danger. A thick wire is sometimes protected by a large fuse, because it is a thick wire, notwithstanding that a small wire is attached to it, unfused; there is always great damage in such cases. It follows, therefore, that wherever the wire changes its size, a fuse should be placed, unless the fuse preceding it is small enough for the smallest wire beyond it. In general, therefore, a fuse should be placed at the beginning of every branch circuit, except as explained.

If only one side of a circuit is protected by fuses, the building is not completely protected, as there are possible cases in which a wire might become overheated, as, for instance, when a heavily-fused wire and a light unfused wire are both grounded or in contact. Fuses should, therefore, always be " double-pole."

It has been suggested to make the fuses of copper wire of a certain number of sizes smaller than the size of the wire to be protected by it. This would be a very good general rule and guide, but the temperature of the fused copper is so very much higher than that of lead alloys, that there would be danger of fire caused by scattering of this melted copper.

Fuses should be marked with the current at which they will fuse, but as such marks are sometimes very unreliable, even with fuses sold by otherwise reliable companies, a careful engineer will always test a sample fuse before using them. Some fuses are marked with the number of lamps normally supplied by them, others with amperes, others with the fusing current, etc.; unless it is known what such marks mean, it is not safe to trust them.

WIRING FORMULÆ. THEIR DEDUCTION AND USE.

When a current passes through a wire there is a gradual loss of voltage along the whole length of the wire. This loss, from Ohm's law, is equal to the product of the current and the resistance, that is,

$$E = C R$$

Now, the resistance of a wire is equal to

$$R = \frac{L}{d^2} 10.61^1$$

in which R is in legal ohms, at about 75° to 80° F.; L is the length of the wire in feet, d is the diameter in mils or d^2 the cross-section in circular mils.

From these two formulæ it follows that

$$E = \frac{10.61\ C\ L}{d^2}$$

from which the loss in volts can be determined for any current, length and diameter of wire. As the circuit is usually a loop or return circuit, it is simpler to use the distance, represented by D which is equal to $\frac{1}{2} L$. Furthermore, as the loss is usually known, while the diameter is that which is required, the formula reduces to the form

$$d^2 = \frac{21.21\ C\ D}{E}$$

in which D is the distance in feet from the dynamo to the lamps or motor, and E is the loss in volts.

For arc light circuits this formula is in its simplest form, and for motor circuits also, after having first determined the current C, which is equal to 746 times the horse-power divided by the voltage of the motor, or which may be found from the tables of horse-power equivalents in the back of this book, see pages 36 and 37.

---

[1] This constant is in accordance with the new Matthiessen standard suggested by the Committee of the American Institute of Electrical Engineers.

For incandescent lighting this formula may be still further simplified by substituting the number of lamps n for the current C, in which case it is necessary to introduce the constant c, which is the current required by one lamp. This is usually multiplied once for all by 21.21, giving what is generally termed the "constant" for calculating the leads for that lamp. The formula then becomes

$$d^2 = n\ D \times \frac{(21.21 \times c)}{E}$$

in which the quantity in parentheses is the "constant" calculated once for all. This constant is then divided by the actual loss in *volts*, E (not in per cent.), which gives a new constant, but for that loss only.

The calculation is therefore as follows : *Multiply the number of lamps by the distance in feet and by the constant* (which constant has first been divided by the loss in volts). The answer is the cross-section in circular mils. From a table (see page 23) find what gauge number this corresponds to, or from a table of squares or square roots find the diameter in mils of which this is the square.

If lamps of different candle-power (and therefore of different currents) are used together, it is best to reduce them all to the equivalent in one size, or else find the total current in amperes, and use the original formulæ in which the current is used instead of the number of lamps.

The loss is often given in per cent. instead of in volts. To find what this is in volts, it is necessary merely to multiply the voltage of the lamp, V, by the per cent. (in whole numbers, thus, 2 per cent.) and divide by 100. Or to bring this all into the formula gives

$$d^2 = n\ D\ \frac{(2121. \times c)}{\%\ V}$$

in which V is the voltage of the lamps, and $\%$ is the loss in per cent. (in units, thus, 2).

Instead of giving the cross-section in circular mils, namely, d^2, the formula might be made to give it in square mils, but the very good practice of using circular mils instead of square mils has become so universal and is so much simpler, that the other is no longer to be recommended. To change the above formulæ so as to give the answer in square mils instead of circular mils, multiply the numerical constant by .7854, and change d^2 to a.

From the above explanation regarding the "constant" anyone

will be able to calculate the constant for any make of lamp. It is always best to calculate this, unless one is very sure what the constant given by the makers means. To determine the constant it is necessary to have the current required for one lamp; whenever possible, it is best to measure this one's self for a batch of 10 or 100 lamps, as the figures given by the makers are sometimes considerably below their true values.

TABLES.

TABLES OF WIRE GAUGES.

Tables giving the diameters and cross-sections of different wire-gauge numbers are usually given separately, or, if together, they usually give approximate equivalents only. As the latter is often insufficient, the accompanying table has been arranged to give in the order of their size all the values for each of the principal American and European gauges. All the approximate equivalents may, therefore, be readily found by mere inspection, while the degree of approximation may be seen directly from their cross-sections or diameters. It therefore forms a complete and combined set of all the gauges used in practice.

The tables usually published often give only approximate diameters and cross-sections, and some of them contain a number of errors. The accompanying table has, therefore, been calculated from the original correct data. It may not be generally known that the tables of B. & S. gauges, as usually published, contain a number of errors which were apparently copied from an incorrect original, and have been acknowledged to be errors by the origi-nators. The corrected values have been used in this table.

In connection with the B. & S. gauge, it may be added here that it follows a regular law, each cross-section being a certain per cent. smaller than the one before. It may not be generally known that with every three sizes the cross-section is doubled approxi-mately. Thus, No. 4, for instance, is very nearly twice as large in cross-section as No. 7 and half as large as No. 1. The error is only one-quarter of 1 per cent. This rule applies to the whole range of the gauge.

The accompanying table may be used also for converting diam-eters into areas, millimetres into mils, diameters of the one into areas of the other units, etc., and *vice versa*.

TABLES OF WIRE GAUGES.
American and European.
WITH CROSS-SECTIONS AND DIAMETERS
Arranged for Comparison and Reduction.

GAUGES AND SCALES.							CROSS-SECTION.			DIAMETER.	
Vulgar Fractions of an Inch.	Millimeter Scale. (Diam. in Millimeters.)	Decimal Scale. (Diam. in Mils.)	Edison Gauge.	Birmingham, or Stubs (Holzapffel) or Old English Standard Gauge. B. W. G.	New British, or Standard Gauge (March 1st, 1884).	American or B. & S. Gauge.	Circular Mils. ($-d^2$) (1 Circular Mil — .7854 Square Mils.)	Square Mils. (1 Sq. Inch — 1,000,000. Sq. Mils.)	Square Millimeters. (1 Sq. m. m. — 1550.1 Sq. Mils).	Millimeters. (1 m. m. — 39.3708 Mils.)	Mils. ($-d$). (1 Inch — 1,000. Mils.)
I.	II.	III.	IV.	V.	VI.	VII.	VIII.	IX.	X.	XI.	XII.
1		1000					1000 000.	785 398.	506.69	25.400	1000.0
¾		750					562 500.	441 786.	285.01	19.050	750.00
⅝		625					390 625.	306 796.	197.93	15.875	625.00
			360				360 000.	282 743.	182.41	15.240	600.00
			340				340 006.	267 040.	172.28	14.810	583.10
			320				320 005.	251 332.	162.14	14.365	565.69
			300				300 008.	235 626.	152.01	13.912	547.73
			280				280 010.	219 920.	141.88	13.440	529.16
			260				260 008.	204 210.	131.74	12.952	509.91
½		500			7/0		250 000.	196 350.	126.68	12.700	500.00
			240				240 002.	188 497.	121.61	12.443	489.90
			220				220 008.	172 794.	111.48	11.914	469.05
					6/0		215 296.	169 093.	109.09	11.785	464.00
						0000	211 600.	166 190.	107.22	11.684	460.00
				0000			206 116.	161 883.	104.44	11.531	454.00
		450					202 500.	159 043.	102.61	11.430	450.00
			200				200 008.	157 084.	101.34	11.359	447.22
7/16							191 406.	150 330.	96.96	11.113	437.50
			190				190 000.	149 226.	96.27	11.071	435.89
					5/0		186 624.	146 574.	94.56	10.972	432.00
		425		000			180 625.	141 863.	91.61	10.795	425.00
			180				180 005.	141 376.	91.21	10.776	424.27
			170				170 008.	133 524.	86.14	10.473	412.32
						000	167 805.	131 790.	85.03	10.405	409.64
		400	160		0000		160 000.	125 664.	81.07	10.160	400.00
	10						155 006.	121 740.	78.54	10.000	393.71
			150				150 001.	117 811.	76.00	9.837	387.30
				00			144 400.	113 411.	73.17	9.652	380.00
⅜		375					140 625.	110 450.	71.25	9.525	375.00
			140				140 008.	109 958.	70.94	9.504	374.17
	9.5						139 893.	109 858.	70.88	9.500	374.02
					000		138 384.	108 667.	70.12	9.448	372.00
						00	133 079.	104 518.	67.43	9.266	364.80
			130				130 004.	102 105.	65.87	9.158	360.56
	9.						125 555.	98 588.	63.62	9.000	354.34
		350					122 500.	96 211.	62.07	8.890	350.00
					00		121 104.	95 115.	61.37	8.839	348.00
			120				120 007.	94 253.	60.81	8.799	346.42
11/32							118 164.	92 810.	59.87	8.731	343.75
		340		0			115 600.	90 792.	58.57	8.636	340.00
	8.5						111 992.	87 958.	56.75	8.500	334.65
			110				110 005.	86 398.	55.74	8.424	331.67
		325					105 625.	82 958.	53.52	8.255	325.00
						0	105 534.	82 887.	53.47	8.251	324.96
					0		104 976.	82 448.	53.19	8.229	324.00
			100				100 001.	78 541.	50.67	8.032	316.23
	8.						99 204.	77 914.	50.27	8.000	314.97
5/16							97 656.	76 699.	49.48	7.937	312.50
			95				95 005.	74 617.	48.14	7.829	308.23
		300	90	1	1		90 000.	70 686.	45.60	7.620	300.00
	7.5						87 191.	68 479.	44.18	7.500	295.28
			85				85 001.	66 760.	43.07	7.405	291.55
						1	83 694.	65 732.	42.41	7.348	289.30
				2			80 656.	63 347.	40.87	7.213	284.00
			80				80 004.	62 835.	40.54	7.184	282.85
9/32							79 102.	62 130.	40.80	7.144	281.25
					2		76 176.	59 826.	38.59	7.010	276.00
	7.						75 953.	59 653.	38.48	7.000	275.60
		275					75 625.	59 390.	38.32	6.985	275.00
			75				75 005.	58 908.	38.00	6.956	273.87

GAUGES AND SCALES.							CROSS-SECTION.			DIAMETER.	
Fractions.	Millimeters.	Decimal.	Edison.	B. W. G.	British.	B. & S.	Circular Mils.	Square Mils.	Square Millimeters.	Millimeters.	Mils.
I.	II.	III.	IV.	V.	VI.	VII.	VIII.	IX.	X.	XI.	XII.
			70				70 003.	54 980.	35.47	6.720	264.58
				3			67 081.	52 685.	33.99	6.578	259.00
						2	66 373.	52 128.	33.63	6.544	257.63
	6.5						65 490.	51 436.	33.18	6.500	255.91
			65				65 005.	51 055.	32.94	6.476	254.96
					3		63 504.	49 876.	32.18	6.401	252.00
¼		250					62 500.	49 087.	31.67	6.350	250.00
			60				60 001.	47 124.	30.40	6.222	244.95
				4			56 644.	44 488.	28.70	6.045	238.00
	6.						55 802.	43 827.	28.27	6.000	236.23
			55				55 004.	43 200.	27.87	5.957	234.53
15/64							54 982.	43 143.	27.83	5.952	234.38
					4		53 824.	42 273.	27.27	5.893	232.00
						3	52 634.	41 339.	26.67	5.827	229.42
		225					50 625.	39 761.	25.65	5.715	225.00
			50				50 001.	39 271.	25.34	5.680	223.61
				5			48 400.	38 013.	24.52	5.588	220.00
7/32							47 852.	37 580.	24.25	5.556	218.75
	5.5						46 889.	36 827.	23.76	5.500	216.54
			45				45 003.	35 346.	22.80	5.388	212.14
					5		44 944.	35 299.	22.77	5.385	212.00
						4	41 743.	32 784.	21.15	5.189	204.31
13/64							41 260.	32 405.	20.91	5.159	203.13
				6			41 209.	32 365.	20.88	5.156	203.00
		200	40				40 000.	31 416.	20.27	5.080	200.00
	5.						38 752.	30 435.	19.63	5.000	196.85
					6		36 864.	28 953.	18.68	4.877	192.00
		190					36 100.	28 350.	18.32	4.826	190.00
	4.8						35 713.	28 055.	18.10	4.800	188.98
3/16							35 156.	27 610.	17.81	4.762	187.50
			35				35 002.	27 491.	17.44	4.752	187.09
						5	33 102.	25 999.	16.77	4.621	181.94
	4.6						32 799.	25 760.	16.62	4.600	181.11
		180		7			32 400.	25 447.	16.42	4.572	180.00
	4.5						31 389.	24 647.	15.90	4.500	177.17
					7		30 976.	24 328.	15.69	4.470	176.00
	4.4						30 009.	23 569.	15.21	4.440	173.23
			30				30 002.	23 563.	15.20	4.400	173.21
11/64							29 541.	23 203.	14.97	4.366	171.88
		170					28 900.	22 698.	14.64	4.318	170.00
	4.2						27 343.	21 475.	13.85	4.200	165.36
				8			27 225.	21 382.	13.79	4.191	165.00
						6	26 250.	20 618.	13.30	4.115	162.02
		160			8		25 600.	20 106.	12.97	4.064	160.00
			25				25 002.	19 636.	12.67	4.016	158.12
	4.						24 801.	19 479.	12.57	4.000	157.48
5/32							24 414.	19 170.	12.37	3.969	156.25
		150					22 500.	17 671.	11.40	3.810	150.00
	3.8						22 383.	17 579.	11.34	3.800	149.61
				9			21 904.	17 203.	11.10	3.759	148.00
						7	20 820.	16 351.	10.55	3.665	144.29
					9		20 736.	16 286.	10.51	3.658	144.00
	3.6						20 089.	15 778.	10.18	3.600	141.74
			20				20 002.	15 710.	10.14	3.592	141.43
9/64							19 776.	15 532.	10.02	3.572	140.63
		140					19 600.	15 394.	9.931	3.556	140.00
	3.5						18 988.	14 913.	9.621	3.500	137.80
				10			17 956.	14 103.	9.098	3.403	134.00
	3.4						17 919.	14 073.	9.079	3.400	133.86
		130					16 900.	13 273.	8.563	3.302	130.00
						8	16 510.	12 967.	8.365	3.264	128.49
					10		16 384.	12 868.	8.302	3.251	128.00
	3.2						15 873.	12 466.	8.042	3.200	125.99
⅛							15 625.	12 270.	7.917	3.175	125.00
			15				15 001.	11 782.	7.601	3.111	122.48
		120		11			14 400.	11 310.	7.296	3.048	120.00
	3.						13 951.	10 954.	7.069	3.000	118.11
					11		13 456.	10 568.	6.818	2.946	116.00
						9	13 092.	10 283.	6.634	2.908	114.42
	2.8						12 152.	9 545.	6.158	2.800	110.24
		110					12 100.	9 503.	6.131	2.794	110.00
			12				12 001.	9 426.	6.081	2.783	109.55
7/64							11 963.	9 395.	6.061	2.778	109.38
				12			11 881.	9 331.	6.020	2.769	109.00
					12		10 816.	8 495.	5.480	2.642	104.00

GAUGES AND SCALES							CROSS-SECTION			DIAMETER	
Fractions	Millimeters	Decimal	Edison	B. W. G.	British	B. & S.	Circular Mils	Square Mils	Square Millimeters	Millimeters	Mils
I.	II.	III.	IV.	V.	VI.	VII.	VIII.	IX.	X.	XI.	XII.
	2.6						10476	8230	5.309	2.600	102.36
						10	10384	8155	5.261	2.588	101.90
		100					10000	7854	5.067	2.540	100.00
	2.5						9688	7609	4.909	2.500	98.43
		95		13			9025	7088	4.573	2.413	95.00
	2.4						8928	7012	4.524	2.400	94.49
3/32							8789	6903	4.453	2.381	93.75
					13		8464	6648	4.289	2.337	92.00
						11	8234	6467	4.172	2.305	90.74
		90					8100	6362	4.104	2.286	90.00
			8				8001	6284	4.054	2.272	89.45
	2.2						7502	5892	3.801	2.200	86.62
		85					7225	5675	3.664	2.160	85.00
				14			6889	5411	3.491	2.108	83.00
						12	6530	5129	3.309	2.053	80.81
		80			14		6400	5027	3.243	2.032	80.00
	2.0						6200	4870	3.142	2.000	78.74
5/64							6104	4793	3.093	1.985	78.13
		75					5625	4418	2.850	1.905	75.00
	1.9						5596	4395	2.835	1.900	74.81
				15	15		5184	4072	2.627	1.829	72.00
						13	5179	4067	2.624	1.828	71.96
	1.8						5022	3944	2.545	1.800	70.87
			5				5001	3928	2.534	1.796	70.72
		70					4900	3848	2.483	1.778	70.00
	1.7						4480	3518	2.271	1.700	66.93
		65		16			4225	3318	2.141	1.651	65.00
						14	4107	3225	2.081	1.628	64.08
					16		4096	3217	2.075	1.626	64.00
	1.6						3968	3115	2.011	1.600	62.99
1/16							3906	3068	1.979	1.588	62.50
		60					3600	2827	1.824	1.524	60.00
	1.5						3488	2739	1.767	1.500	59.06
				17			3364	2642	1.705	1.473	58.00
						15	3257	2558	1.650	1.450	57.07
					17		3136	2463	1.589	1.422	56.00
	1.4						3038	2384	1.539	1.400	55.12
		55	3				3025	2376	1.533	1.397	55.00
						16	3001	2357	1.521	1.391	54.78
	1.3						2620	2057	1.327	1.300	51.18
		50				16	2583	2029	1.309	1.291	50.82
							2500	1964	1.267	1.270	50.00
				18			2401	1886	1.217	1.245	49.00
					18		2304	1810	1.167	1.219	48.00
	1.2						2232	1753	1.131	1.200	47.25
3/64							2197	1726	1.113	1.191	46.88
						17	2048	1609	1.038	1.150	45.26
		45					2025	1590	1.026	1.143	45.00
	1.1						1875	1473	.9509	1.100	43.31
				19			1764	1385	.8938	1.067	42.00
						18	1624	1276	.8230	1.024	40.30
		40			19		1600	1257	.8107	1.016	40.00
	1.0						1550	1217	.7854	1.000	39.37
					20		1296	1018	.6567	.9144	36.00
						19	1288	1012	.6527	.9116	35.89
	.9						1256	985.9	.6362	.9000	35.43
		35		20			1225	962.1	.6207	.8890	35.00
				21	21		1024	804.2	.5188	.8128	32.00
						20	1022	802.3	.5176	.8118	31.96
	.8						992.0	779.3	.5027	.8000	31.50
1/32							976.6	767.0	.4948	.7937	31.25
		30					900.0	706.9	.4560	.7620	30.00
						21	810.1	636.3	.4105	.7229	28.46
		28		22	22		784.0	615.8	.3972	.7112	28.00
	.7						759.5	596.5	.3848	.7000	27.56
		26					676.0	530.9	.3425	.6604	26.00
						22	642.5	504.6	.3255	.6438	25.35
		25		23			625.0	490.9	.3167	.6350	25.00
		24			23		576.0	452.4	.2919	.6096	24.00
	.6						568.0	438.3	.2827	.6000	23.62
						23	509.5	400.2	.2581	.5733	22.57
		22		24	24		484.0	380.1	.2451	.5588	22.00
						24	404.1	317.3	.2047	.5106	20.10
		20		25	25		400.0	314.2	.2027	.5080	20.00
	.5						387.5	304.4	.1963	.5000	19.69

| GAUGES AND SCALES | | | | | | | CROSS-SECTION | | | DIAMETER | |
| Fractions | Millimeters | Decimal | Edison | B. W. G. | British | B. & S. | Circular Mils | Square Mils | Square Millimeters | Millimeters | Mils |
I.	II.	III.	IV.	V.	VI.	VII.	VIII.	IX.	X.	XI.	XII.
		19					361.0	283.5	.1832	.4826	19.00
		18		26	26		324.0	254.5	.1642	.4572	18.00
						25	320.4	251.7	.1624	.4547	17.90
	.45						313.9	246.5	.1590	.4500	17.72
		17					289.0	227.0	.1464	.4318	17.00
					27		269.0	211.2	.1363	.4166	16.40
		16		27			256.0	201.1	.1297	.4064	16.00
						26	254.1	199.6	.1288	.4049	15.94
	.4						248.0	194.8	.1257	.4000	15.75
1/64							244.1	191.8	.1237	.3969	15.63
		15					225.0	176.7	.1140	.3810	15.00
					28		219.0	172.0	.1110	.3759	14.80
						27	201.5	158.3	.1021	.3606	14.20
		14		28			196.0	153.9	.099 31	.3556	14.00
	.35						189.9	149.1	.096 21	.3500	13.78
					29		185.0	145.3	.093 72	.3454	13.60
		13		29			169.0	132.7	.085 63	.3302	13.00
						28	159.8	125.5	.080 97	.3211	12.64
					30		153.8	120.8	.077 92	.3150	12.40
		12		30			144.0	113.1	.072 96	.3048	12.00
	.3						139.5	109.5	.070 69	.3000	11.81
					31		134.6	105.7	.068 18	.2946	11.60
						29	126.7	99.53	.064 21	.2859	11.26
	.28						121.5	95.45	.061 58	.2800	11.02
		11					121.0	95.03	.060 87	.2794	11.00
					32		116.6	91.61	.059 10	.2743	10.80
	.26						104.8	82.30	.053 03	.2600	10.24
						30	100.5	78.94	.050 92	.2546	10.03
		10		31	33		100.0	78.54	.050 67	.2540	10.00
	.24						89.28	70.12	.045 24	.2400	9.449
					34		84.64	66.48	.042 89	.2337	9.200
		9		32			81.00	63.62	.041 04	.2286	9.000
						31	79.70	62.60	.040 39	.2268	8.928
	.22						75.02	58.90	.038 01	.2200	8.662
					35		70.56	55.42	.035 75	.2134	8.400
		8		33			64.00	50.27	.032 43	.2032	8.000
						32	63.20	49.64	.032 03	.2019	7.950
	.20						62.00	48.70	.031 42	.2000	7.874
					36		57.76	45.36	.029 26	.1930	7.600
	.18						50.22	39.44	.025 45	.1800	7.087
						33	50.13	39.37	.025 39	.1798	7.080
		7		34			49.00	38.48	.024 83	.1778	7.000
					37		46.24	36.32	.023 43	.1727	6.800
						34	39.75	31.22	.020 14	.1601	6.305
	.16						39.68	31.15	.020 11	.1600	6.299
		6			38		36.00	28.27	.018 24	.1524	6.000
						35	31.53	24.76	.015 97	.1428	5.615
	.14						30.38	23.84	.015 39	.1400	5.512
					39		27.04	21.24	.013 70	.1321	5.200
		5		35		36	25.00	19.64	.012 67	.1270	5.000
					40		23.04	18.10	.011 67	.1219	4.800
	.12						22.32	17.53	.011 31	.1200	4.725
						37	19.83	15.57	.010 05	.1131	4.453
					41		19.36	15.21	.0098 09	.1117	4.400
	.11						18.75	14.73	.0095 03	.1100	4.331
		4		36	42		16.00	12.57	.0081 07	.1016	4.000
						38	15.72	12.35	.0079 67	.1007	3.965
	.10						15.50	12.17	.0078 54	.1000	3.937
					43		12.96	10.18	.0065 61	.0914	3.600
	.09						12.56	9.859	.0063 62	.0900	3.543
						39	12.47	9.793	.0063 18	.0897	3.531
					44		10.24	8.042	.0051 91	.0813	3.200
	.08						9.920	7.793	.0050 27	.0800	3.150
						40	9.888	7.766	.0050 10	.0799	3.145
		3					9.000	7.069	.0045 60	.0762	3.000
					45	41	7.840	6.158	.0039 72	.0711	2.800
	.07						7.595	5.965	.0038 48	.0700	2.756
					46		5.760	4.524	.0029 18	.0610	2.400
	.06						5.580	4.383	.0028 27	.0600	2.362
		2			47		4.000	3.142	.0020 27	.0508	2.000
	.05						3.875	3.044	.0019 63	.0500	1.969
					48		2.560	2.011	.0012 97	.0406	1.600
					49		1.440	1.131	.0007 30	.0305	1.200
		1			50		1.000	.7854	.0005 07	.0254	1.000

COMPOUNDED WIRES OF LARGE CROSS-SECTION.

In wiring, it is sometimes necessary to use wires larger than No. 00, B. & S. gauge. It then becomes necessary to compound the wire, not only because No. 00 is the largest size which it is practicable to lay (unless the wire is stranded), but chiefly because the size wanted does not generally happen to correspond with those of the gauge numbers; and as the length of the wires is often great, a small excess over the required cross-section may signify a considerable increase in the cost. In such cases it is, therefore, often desirable to obtain the closest possible approximation to the required cross-section by the best combination of the sizes in the market.

The table gives every possible combination of the four largest wires which it is practicable to use, namely, Nos. 2, 1, 0 and 00 B. & S. gauge. The combinations are classified in the order of their combined sections. Having given the desired cross-section of a compounded wire, for instance, 400,000 circular mils, look for this size in the second column, then all the possible combinations which approximate this most closely will be found near to it in the adjoining first column. In this case it will be seen from the table that three No. 0 wires and one No. 1 will give it very closely; and there is no other combination which will give it more closely. Furthermore, the values often do not differ very much from each other, thus allowing some choice, which is often desirable. For instance, in this case it will be seen that three No. 00 wires will give practically the same close approximation, and this would require the handling of only one size of wire, which is sometimes greatly to be preferred. Again, the combination just above this one, namely, four No. 1 wires and one No. 2, is also quite close to the desired value; this combination would be preferable if there are many corners and bends, as the wires are smaller.

The largest limit of the cross-sections in this table was taken as 500,000 circular mils, or a little less than four 00 wires. For larger sections, as, for instance, 600,000, select from the table any convenient combination, regardless of cross-section, as, for instance, that of three 00 wires, and subtract its combined section, namely about 400,000, from the 600,000, and then find from the table the best combination to make up this balance of 200,000, as, for instance, one No. 00 and one No. 2 wire.

TABLE OF

COMPOUNDED WIRES OF LARGE SECTION.

A table of all the possible combinations of numbers 00, 0, 1 and 2, B. & S. wires having a combined cross-section of less than 500,000 circular mils.

B. & S. (American) Gauge Numbers.	Combined Cross-section in Circular Mils.	B. & S. (American) Gauge Numbers.	Combined Cross-section in Circular Mils.	B. & S. (American) Gauge Numbers.	Combined Cross-section in Circular Mils.
00-00-00-00	532 316.	00-0-2-2-2	437 732.	0-0-2-2	343 814.
0-0-0-0-0	527 670.	0-2-2-2-2-2	437 399.	0-1-1-2	339 295.
1-1-1-1-1-1	502 164.	00-00-1-1	433 546.	1-1-1-1	334 776.
00-00-1-1-2	499 919.	00-1-1-2-2	433 213.	00-00-2	332 531.
00-1-1-2-2-2	499 586.	1-1-2-2-2-2	432 880.	00-2-2-2	332 198.
1-1-2-2-2-2-2	499 253.	00-0-0-1	427 841.	2-2-2-2-2	331 865.
00-0-0-1-2	494 214.	0-0-1-2-2	427 508.	00-0-1	322 307.
0-0-1-2-2-2	493 881.	0-1-1-1-2	422 989.	0-1-2-2	321 974.
00-0-1-1-1	489 695.	0-0-0-0	422 136.	1-1-1-2	317 455.
0-1-1-1-2-2	489 362.	1-1-1-1-1	418 470.	0-0-0-0	316 602.
0-0-0-0-0-2	488 509.	00-00-1-2	416 225.	00-0-2	304 986.
1-1-1-1-1-2	484 843.	00-1-2-2-2	415 892.	0-2-2-2	304 653.
0-0-0-1-1	483 990.	1-2-2-2-2-2	415 559.	00-1-1	300 467.
00-00-00-1	482 931.	00-0-0-2	410 520.	1-1-2-2	300 134.
00-00-1-2-2	482 598.	0-0-2-2-2	410 187.	0-0-1	294 762.
00-1-2-2-2-2	482 265.	00-0-1-1	406 001.	00-1-2	283 146.
1-2-2-2-2-2-2	481 932.	0-1-1-2-2	405 668.	1-2-2-2	282 813.
00-00-0-0-0	477 226.	1-1-1-1-2	401 140.	0-0-2	277 441.
00-0-0-2-2	476 893.	0-0-0-1	400 296.	0-1-1	272 922.
0-0-2-2-2-2	476 560.	00-00-00	399 237.	00-00	266 153.
00-0-1-1-2	472 374.	00-00-2-2	398 904.	00-2-2	265 825.
0-1-1-2-2-2	472 041.	00-2-2-2-2	396 571.	2-2-2-2	265 492.
00-1-1-1-1	467 855.	2-2-2-2-2-2	396 238.	0-1-2	255 601.
1-1-1-1-2-2	467 522.	00-0-1-2	388 680.	1-1-1	251 082.
0-0-0-1-2	466 669.	0-1-2-2-2	388 347.	00-0	238 613.
00-00-00-2	465 610.	00-1-1-1	384 161.	0-2-2	238 280.
00-00-2-2-2	465 277.	1-1-1-2-2	383 828.	1-1-2	233 761.
00-2-2-2-2-2	464 944.	0-0-0-0	382 075.	00-1	216 773.
2-2-2-2-2-2-2	464 611.	0-0-1-1	378 456.	1-2-2	216 440.
0-0-1-1-1	462 150.	00-00-0	371 692.	0000	211 600.
00-00-0-1	455 386.	00-0-2-2	371 359.	0-0	211 066.
00-0-1-2-2	455 053.	0-2-2-2-2	371 026.	00-2	199 452.
0-1-2-2-2-2	454 720.	00-1-1-2	366 840.	2-2-2	199 119.
00-1-1-1-2	450 534.	1-1-2-2-2	366 507.	0-1	189 228.
1-1-1-2-2-2	450 201.	0-0-1-2	361 135.	0-2	171 907.
00-0-0-0	449 681.	0-1-1-1	356 676.	000	167 805.
0-0-0-2-2	449 348.	00-00-1	349 852.	1-1	167 388.
0-0-1-1-2	444 829.	00-1-2-2	349 519.	1-2	150 067.
0-1-1-1-1	440 310.	1-2-2-2-2	349 186.	00	133 079.
00-00-0-2	438 065.	00-0-0	344 147.	2-2	132 746.

WIRING COMPUTER.

/ TABLE OF

WEIGHT AND RESISTANCE OF COPPER WIRE.

American or B. & S. Wire Gauge.	Decimal Gauge in Mils.	New British Gauge, or Standard Wire Gauge, March, 1884.	Diameter in Mils. (1 mil = .001 inch.)	Cross-section in Circular Mils. (Circ. mil = .7854 sq. mil.)	Cross-section in Square Mils. (1 sq. in. = 1,000,000 sq. mils.)	Pounds per 1000 Feet (Sp. gr. 8.889.)	Feet per Pound.	Ohms per 1000 Feet (1 mil-foot 10.666 legal ohms.)	Ohms per Pound.	Feet per Ohm.	Pounds per Ohm.
...	500.	7/0	500.00	250 000.	196 350.	756.6	1.322	.04242	.0000561	23573.	17836.
...	...	6/0	464.00	215 296.	169 093.	651.6	1.535	.04926	.0000756	20301.	13228.
4/0	...	...	460.00	211 600.	166 190.	640.4	1.562	.05012	.0000783	19958.	12778.
...	450.	...	450.00	202 500.	159 043.	612.9	1.632	.05237	.0000855	19094.	11702.
...	...	5/0	432.00	186 624.	146 574.	564.8	1.770	.05683	.0001006	17598.	9939.
...	425.	...	425.00	180 625.	141 863.	534.2	1.829	.05871	.0001074	17032.	9310.
000	...	...	409.64	167 805.	131 790.	507.9	1.969	.06320	.0001244	15823.	8036.
...	400.	4/0	400.00	160 000.	125 664.	484.2	2.065	.06628	.0001369	15087.	7306.
...	375.	...	375.00	140 625.	110 450.	425.6	2.350	.07542	.0001772	13260.	5643.
...	...	000	372.00	138 384.	108 687.	418.8	2.388	.07664	.0001830	13049.	5465.
00	...	...	364.80	133 079.	104 518.	402.8	2.483	.07969	.0001979	12548.	5054.
...	350.	...	350.00	122 500.	96 211.	370.8	2.697	.08657	.0002335	11551.	4282.
...	...	00	348.00	121 104.	95 115.	366.5	3.728	.08757	.0002389	11419.	4185.
...	325.	...	325.00	105 625.	82 964.	319.7	3.128	.1004	.0003141	9960.	3184.
0	...	...	324.86	105 534.	82 887.	319.4	3.131	.1005	.0003146	9951.	3178.
...	...	0	324.00	104 976.	82 448.	317.7	3.148	.1010	.0003180	9899.	3145.
...	300.	1	300.00	90 000.	70 686.	272.4	3.671	.1178	.0004326	8486.	2312.
1	...	...	289.30	83 694.	65 732.	253.3	3.948	.1267	.0005003	7892.	1999.
...	...	2	276.00	76 176.	59 828.	230.5	4.338	.1392	.0006039	7183.	1656.
...	275.	...	275.00	75 625.	59 390.	228.9	4.369	.1402	.0006127	7131.	1632.
2	...	...	257.63	66 373.	52 128.	200.9	4.978	.1598	.0007955	6258.	1257.
...	...	3	252.00	63 504.	49 876.	192.2	5.203	.1670	.0008689	5988.	1151.
...	250.	...	250.00	62 500.	49 087.	189.2	5.287	.1697	.0008971	5893.	1175.
...	...	4	232.00	53 824.	42 273.	162.9	6.139	.1970	.001210	5075.	826.8
3	...	...	229.42	52 634.	41 339.	159.3	6.278	.2015	.001265	4963.	790.6
...	225.	...	225.00	50 625.	39 761.	153.2	6.527	.2095	.001367	4774.	731.4
...	...	5	212.00	44 944.	35 299.	136.0	7.352	.2360	.001735	4238.	576.4
4	...	...	204.31	41 743.	32 784.	126.3	7.916	.2541	.002011	3936.	497.2
...	200.	...	200.00	40 000.	31 416.	121.1	8.260	.2651	.002190	3772.	456.6
...	...	6	192.00	36 864.	28 963.	111.6	8.963	.2877	.002579	3476.	387.8
5	181.94	...	181.94	33 102.	25 999.	100.2	9.982	.3204	.003198	3121.	312.7
...	180.	...	180.00	32 400.	25 447.	98.06	10.20	.3273	.003338	3055.	299.6
6	...	7	176.00	30 976.	24 328.	93.75	10.67	.3424	.003652	2921.	273.8
...	...	...	162.02	26 250.	20 618.	79.45	12.59	.4040	.005085	2475.	196.7
...	160.	8	160.00	25 600.	20 106.	77.48	12.91	.4143	.005347	2414.	187.0
7	...	...	144.29	20 820.	16 351.	63.01	15.87	.5094	.008065	1963.	123.7
...	...	9	144.00	20 736.	16 286.	62.76	15.93	.5114	.008150	1955.	122.7
...	140.	...	140.00	19 600.	15 394.	59.32	16.86	.5411	.009121	1848.	109.6
...	130.	...	130.00	16 900.	13 273.	51.15	19.55	.6275	.01227	1594.	81.51
8	...	...	128.49	16 510.	12 967.	49.97	20.01	.6424	.01286.	1557.	77.79
...	...	10	128.00	16 384.	12 868.	49.59	20.17	.6473	.01305	1545.	76.80
...	120.	...	120.00	14 400.	11 310.	43.56	22.95	.7365	.01690	1358.	59.18
...	...	11	116.00	13 456.	10 568.	40.73	24.67	.7881	.01935	1269.	51.67
9	...	...	114.42	13 092.	10 283.	39.63	25.24	.8100	.02044	1235.	48.92
...	110.	...	110.00	12 100.	9 503.	36.62	27.31	.8765	.02393	1141.	41.78
10	...	12	104.00	10 816.	8 495.	32.73	30.55	.9805	.02995	1020.	33.39
...	100.	...	101.90	10 384.	8 155.	31.42	31.82	1.021	.03250	979.1	30.77
...	...	...	100.00	10 000.	7 854.	30.27	33.04	1.061	.03504	942.9	28.54
11	...	13	92.000	8 464.	6 648.	25.62	39.04	1.253	.04891	798.1	20.44
...	...	...	90.742	8 234.	6 467.	24.92	40.13	1.288	.05168	776.4	19.35
...	90.	...	90.000	8 100.	6 362.	24.51	40.79	1.309	.05341	763.8	18.72
12	...	...	80.808	6 530.	5 129.	19.76	50.60	1.624	.08218	615.7	12.17
...	80.	14	80.000	6 400.	5 027.	19.37	51.63	1.657	.08555	603.5	11.69
...	...	15	72.000	5 184.	4 072.	15.69	63.74	2.046	.1304	488.8	7.669
13	...	...	71.962	5 179.	4 067.	15.67	63.81	2.048	.1307	488.3	7.653
...	70.	...	70.000	4 900.	3 848.	14.83	67.43	2.164	.1459	462.0	6.852
14	...	...	64.084	4 107.	3 225.	12.43	80.46	2.582	.2078	387.2	4.813
...	...	16	64.000	4 096.	3 217.	12.40	80.67	2.589	.2084	386.2	4.788
...	60.	...	60.000	3 600.	2 827.	10.90	91.78	2.946	.2704	339.5	3.698
15	...	...	57.068	3 257.	2 558.	9.857	101.5	3.256	.3304	307.1	3.027
...	...	17	56.000	3 136.	2 463.	9.491	105.4	3.382	.3646	295.7	2.807
16	...	...	50.821	2 583.	2 029.	7.817	127.9	4.106	.5253	243.5	1.904
...	50.	...	50.000	2 500.	1 964.	7.566	132.2	4.242	.5607	235.7	1.784
...	...	18	48.000	2 304.	1 810.	6.973	143.4	4.603	.6601	217.3	1.515
17	...	...	45.257	2 048.	1 609.	6.199	161.3	5.178	.8353	193.1	1.197

According to the Matthiessen Standard suggested by the Committee of the Amer. Inst. of Elect. Eng., these resistances are for pure copper wire at 78¼° F.

B. & S. Gauge.	Decimal Gauge.	New British Gauge.	Diam. in Mils.	Cross-section in Circular Mils.	Cross-section in Square Mils.	Pounds per 1000 Feet.	Feet per Pound.	Ohms per 1000 Feet.	Ohms per Pound.	Feet per Ohm.	Pounds per Ohm.
18	45.	...	45.000	2 025.	1 590.	6.129	163.2	5.237	.8545	190.9	1.170
...	40.	19	40.303	1 624.	1 276.	4.916	203.4	6.529	1.325	153.2	.7529
...	...	...	40.000	1 600.	1 257.	4.842	206.5	6.628	1.369	150.9	.7306
...	...	20	36.000	1 296.	1.018.	3.922	255.0	8.183	2.086	122.2	.4793
19	...	...	35.891	1 288.	1.012	3.899	256.5	8.233	2.112	121.5	.4735
...	35.	...	35.000	1 225.	962.1	3.708	269.7	8.657	2.335	115.5	.4282
...	...	21	32.000	1 024.	804.2	3.099	322.7	10.36	3.342	96.56	.2992
20	...	...	31.961	1 022.	802.3	3.092	323.5	10.38	3.358	96.33	.2978
...	30.	...	30.000	900.0	706.9	2.724	367.1	11.78	4.326	84.86	.2312
21	...	...	28.462	810.1	636.3	2.452	407.9	13.09*	5.339	76.39	.1873
...	28.	22	28.000	784.0	615.8	2.379	421.4	13.53	5.701	73.93	.1754
...	26.	...	26.000	676.0	530.9	2.046	488.8	15.69	7.668	63.74	.1304
22	...	...	25.347	642.5	504.6	1.944	514.3	16.51	8.490	60.58	.1178
...	24.	23	24.000	576.0	452.4	1.743	573.6	18.41	10.56	54.31	.09458
23	...	...	22.572	509.5	400.2	1.542	648.5	20.82	13.50	48.04	.07408
...	22.	24	22.000	484.0	380.1	1.465	682.7	21.91	14.96	45.64	.06685
24	...	...	20.101	404.1	317.3	1.223	817.8	26.25	21.47	38.10	.04659
...	20.	25	20.000	400.0	314.2	1.211	826.0	26.51	21.90	37.72	.04566
...	18.	26	18.000	324.0	254.5	.9806	1020.	32.73	33.38	30.55	.02996
25	...	...	17.900	320.4	251.7	.9697	1031.	33.10	34.13	30.21	.02930
...	...	27	16.400	269.0	211.2	.8140	1229.	39.43	48.44	25.36	.02064
...	16.	...	16.000	256.0	201.1	.7748	1291.	41.43	53.47	24.14	.01870
26	...	...	15.941	254.1	199.6	.7690	1300.	41.74	54.27	23.96	.01843
...	15.	...	15.000	225.0	176.7	.6810	1468.	47.13	69.22	21.22	.01445
...	...	28	14.800	219.0	172.0	.6629	1508.	48.42	73.04	20.65	.01369
27	...	...	14.195	201.5	158.3	.6099	1640.	52.63	86.29	19.00	.01159
...	14.	...	14.000	196.0	153.9	.5932	1686.	54.11	91.21	18.48	.01096
...	...	29	13.600	185.0	145.3	.5598	1786.	57.34	102.4	17.44	.009763
...	13.	...	13.000	169.0	132.7	.5115	1955.	62.75	122.7	15.94	.008151
28	...	...	12.641	159.8	125.5	.4836	2068.	66.86	137.2	15.07	.007288
...	...	90	12.400	153.8	120.8	.4654	2149.	68.97	148.2	14.50	.006747
...	12.	...	12.000	144.0	113.1	.4358	2296.	73.65	169.0	13.58	.005918
...	...	31	11.600	134.6	105.7	.4073	2456.	78.81	193.5	12.69	.005167
29	...	...	11.258	126.7	99.54	.3836	2607.	83.68	218.2	11.95	.004584
...	11.	...	11.000	121.0	95.03	.3662	2731.	87.65	239.3	11.41	.004178
...	...	32	10.800	116.6	91.61	.3530	2833.	90.92	257.6	11.00	.003883
30	...	...	10.025	100.5	78.94	.3042	3288.	105.5	346.9	9.477	.002883
...	10.	33	10.000	100.0	78.54	.3027	3304.	106.1	350.4	9.429	.002854
...	...	34	9.2000	84.64	66.48	.2562	3904.	125.3	489.1	7.981	.002044
...	9.	...	9.0000	81.00	63.62	.2451	4079.	130.9	534.1	7.638	.001872
31	...	...	8.9277	79.70	62.60	.2412	4146.	133.1	551.6	7.515	.001813
...	8.	35	8.4000	70.56	55.42	.2136	4683.	150.3	703.8	6.653	.001421
...	...	...	8.0000	64.00	50.27	.1937	5163.	165.7	855.5	6.035	.001169
32	...	36	7.9503	63.20	49.64	.1913	5228.	167.8	877.1	5.960	.001140
...	...	...	7.6000	57.76	45.36	.1748	5720.	183.6	1050.	5.446	.0009521
33	...	...	7.0800	50.13	39.37	.1517	6592.	211.6	1395.	4.727	.0007170
...	7.	...	7.0000	49.00	38.48	.1483	6743.	216.4	1459.	4.620	.0006852
...	...	37	6.8000	46.24	36.32	.1399	7146.	224.1	1639.	4.360	.0006102
34	...	...	6.3049	39.75	31.22	.1203	8312.	266.8	2218.	3.748	.0004510
...	6.	38	6.0000	36.00	28.27	.1090	9178.	294.6	2704.	3.395	.0003698
35	...	...	5.6147	31.53	24.76	.09541	10482.	336.4	3526.	2.973	.0002836
...	...	39	5.2000	27.04	21.24	.08184	12220.	392.2	4792.	2.550	.0002087
36	5.	...	5.0000	25.00	19.64	.07566	13217.	424.2	5607.	2.357	.0001784
...	...	40	4.8000	23.04	18.10	.06973	14341.	460.3	6601.	2.173	.0001515
37	...	...	4.4526	19.83	15.57	.06000	16666.	534.9	8915.	1.869	.0001122
...	4.	41	4.4000	19.36	15.21	.05859	17067.	547.8	9349.	1.826	.0001070
...	...	42	4.0000	16.00	12.57	.04842	20651.	662.8	13688.	1.509	.0000731
38	...	...	3.9652	15.72	12.35	.04758	21015.	674.5	14175.	1.489	.0000706
...	...	43	3.6000	12.96	10.18	.03922	25495.	818.3	20853.	1.222	.0000479
39	...	...	3.5311	12.47	9.793	.03774	26500.	850.6	22540.	1.176	.0000444
...	...	44	3.2000	10.24	8.042	.03099	32267.	1036.	33418.	.9656	.0000299
40	3.	...	3.1445	9.888	7.765	.02993	33416.	1073.	35841.	.9324	.0000279
...	...	...	3.0000	9.000	7.069	.02724	36713.	1178.	43260.	.8486	.0000231
...	...	45	2.8000	7.840	6.158	.02373	42146.	1353.	57009.	.7393	.0000175
...	...	46	2.4000	5.760	4.524	.01743	57364.	1841.	105620.	.5431	.0000095
...	2.	...	2.0000	4.000	3.142	.01211	82604.	2651.	219010.	.3772	.0000046
...	...	47	1.6000	2.560	2.011	.00775	129068.	4143.	534690.	.2414	.0000019
...	...	48	1.2000	1.440	1.131	.00496	229456.	7365.	1689900.	.1358	.0000006
...	1.	50	1.0000	1.000	.7854	.00303	330416.	10605.	3504100.	.0943	.0000003
...	...	...	35.682	1273.	1000.	3.853	259.5	8.329	2.162	120.1	.4628
...	...	...	19.177	330.4	259.5	1.000	1000.	32.10	32.10	31.16	.03118
...	...	...	574.92	330418.	259510.	1000.	1.000	.03210	.0003210	31156.	31156.
...	...	...	102.98	10605.	8329.	32.10	31.16	1.000	.03116	1000.	32.10
...	...	...	43.266	1872.	1470.	5.665	176.5	1.000	.03210	176.5	32.10
...	...	...	3.2566	10.61	8.329	.03210	31156.	1000.	31156.	1.000	.0000321
...	...	...	43.266	1872.	1470.	5.665	176.5	5.665	1.000	176.5	1.000

TABLE OF
TEMPERATURE CORRECTIONS FOR COPPER WIRE.

Instead of using the usual formula for correcting the resistance of copper wire for temperature, the calculation may be very much simplified by finding the mil-foot resistance K in the first column of the accompanying table, corresponding to the given temperature, and using the simple formula $R = \dfrac{L}{d^2} K$, in which R is the required resistance in legal ohms at the given temperature; L is the length in feet; d is the diameter of the wire in mils, or d^2 the cross-section in circular mils; and K is the mil-foot resistance taken from the table. As this constant contains only two digits, one of which is unity, the calculation is a very simple one.

This table is based on the Matthiessen standard suggested by the Committee of the American Institute of Electrical Engineers, namely 9.612 legal ohms for a mil-foot at 0° C.

Resistance per Mil-foot in Legal Ohms. K.	Temperature in Fahrenheit Degrees.	Temperature in Centigrade Degrees.	Resistance per Mil-foot in Legal Ohms. K.	Temperature in Fahrenheit Degrees.	Temperature in Centigrade Degrees.
10.00	50.47	10.26	10.80	86.90	30.50
10.10	55.15	12.86	10.90	91.31	32.95
10.20	59.79	15.44	11.00	95.69	35.38
10.30	64.40	18.00	11.10	100.04	37.80
10.40	68.97	20.54	11.20	104.36	40.20
10.50	73.51	23.06	11.30	108.64	42.58
10.60	78.01	25.56	11.40	112.90	44.95
10.70	82.47	28.04	11.50	117.14	47.30

WEIGHT OF INSULATED WIRE FOR WIRING.

FOR COMPUTING THE COST WHEN MAKERS GIVE THE PRICES PER POUND INSTEAD OF PER 100. FEET.

WEIGHTS IN POUNDS PER 100. FEET.

B. & S. Wire Gauge Numbers.		American Electrical Works. Underwriters Braided Electric Light Line Wire.	American Electrical Works. Weather-proof Braided Electric Light Line Wire.	Holmes, Booth and Haydens. K. K. Triple-braided.	A. F. Moore. Underwriters.	A. F. Moore. Weather-proof.	A. F. Moore. Fire and Weather-proof.	N. Y. Insulated Wire Co. Competition Line Wire.	N. Y. Insulated Wire Co. Other Wires.	Okonite Electric Light Line Wires. Plain Insulation.	Okonite Electric Light Line Wires. Braided Insulation.	Simplex. T Z R Weather-proof.	Simplex. Caoutchouc, Plain Rubber.	Simplex. Caoutchouc with Protective Braids.	B. & S. Wire Gauge Numbers.
0000	Solid	...	...	70.6	75.0	73.0	88.7	...	...	93.8	99.0	74.6	78.1	92.6	0000
000	Solid	...	...	60.0	65.0	58.7	65.5	...	...	69.2	71.4	60.6	63.3	80.9	000
00	Solid	45.0	42.5	50.0	44.0	42.3	51.6	...	...	56.4	60.0	47.7	50.1	67.2	00
0	Solid	35.0	33.0	40.0	35.0	33.2	41.7	40.0	...	43.7	47.3	38.2	41.8	57.0	0
1	Solid	29.0	27.0	31.6	28.4	27.0	33.4	33.3	...	34.5	36.7	31.1	31.4	37.2	1
1	Stranded	...	...	...	...	...	...	...	...	36.2	38.7	...	33.0	41.2	1
2	Solid	24.0	20.4	27.9	23.5	22.4	27.7	28.6	...	28.2	29.7	2.38	25.5	29.0	2
2	Stranded	...	...	...	...	...	...	...	...	30.0	32.5	...	26.5	31.9	2
3	Solid	19.5	17.7	24.0	19.0	18.0	22.3	21.1	...	20.6	22.0	19.2	20.3	24.0	3
3	Stranded	...	...	...	...	...	...	...	...	24.0	25.7	...	22.1	25.8	3
4	Solid	15.5	14.0	15.8	15.5	14.7	18.2	18.2	...	17.0	18.4	15.3	16.7	18.8	4
4	Stranded	...	...	...	...	...	...	...	...	20.1	21.4	...	17.1	19.1	4
5	Solid	12.5	11.0	12.9	12.5	11.9	17.4	14.3	...	...	...	12.6	13.6	16.3	5
5	Stranded	...	...	...	...	...	...	...	...	16.3	17.7	...	13.8	...	5
6	Solid	10.5	9.5	10.9	10.2	9.7	12.0	11.1	...	11.6	12.5	10.4	10.7	12.7	6
6	Stranded	...	...	...	...	...	...	...	...	13.6	15.0	...	11.7	13.5	6
7	Solid	8.1	7.3	8.3	8.1	7.7	10.3	...	...	...	...	8.9	8.5	10.2	7
7	Stranded	...	...	...	...	...	...	...	...	10.6	11.5	...	...	...	7
8	Solid	7.3	6.5	7.1	6.9	6.6	8.9	7.2	...	7.0	7.6	7.6	7.0	8.2	8
8	Stranded	...	...	...	...	...	...	...	...	8.8	9.6	...	7.4	8.7	8
9	Solid	5.5	4.9	5.5	5.4	5.1	6.8	...	...	5.6	6.1	6.4	5.9	7.0	9
10	Solid	5.0	4.5	5.2	4.7	4.5	6.0	...	...	5.2	5.6	5.3	4.6	5.7	10
11	Solid	4.0	3.5	3.95	...	...	...	...	...	...	...	...	...	...	11
12	Solid	2.9	2.5	3.40	2.65	2.6	3.5	...	...	3.3	3.8	3.7	3.1	4.2	12
13	Solid	2.4	2.1	...	...	...	...	...	...	...	...	...	...	...	13
14	Solid	2.1	1.8	2.27	2.00	1.9	2.7	...	...	2.4	2.7	2.4	2.1	2.9	14
15	Solid	1.7	1.5	...	...	...	...	...	...	...	...	...	...	...	15
16	Solid	1.3	1.3	1.89	1.30	1.25	1.8	...	...	1.55	1.66	1.8	1.4	1.9	16
17	Solid	1.2	1.2	...	...	...	...	...	...	...	...	...	...	...	17
18	Solid	1.0	1.0	1.50	1.05	1.00	1.4	...	...	1.10	1.38	1.5	1.0	1.4	18
19	Solid	.90	.90	...	.90	.85	1.24	...	...	.95	1.21	...	...	...	19
20	Solid	.85	.85	...	.85	.80	1.17	...	...	...	...	...	...	...	20

Ordered and sold by the 100 feet and not by the pound.

TABLE OF HEATING LIMITS

OR

MAXIMUM SAFE CARRYING CAPACITY

OF INSULATED WIRES.

These numbers were calculated from the formula given by the Edison Company on their standard tables, namely : max. amp. $=$

$$\sqrt[4]{\left(\frac{circ.\ mils}{104.}\right)^3}$$ which reduces to the more convenient form:

$$.031 \sqrt[2]{diam.^3}$$

The numbers are only approximate, as they depend on the nature of the surroundings of the wire, thickness of insulation, etc. The temperature given with the formula is 50° C. or 122° F.

B. & S. Gauge Number.	Amperes.	Greatest number of LAMPS of the following different currents per lamp: (For the THREE-WIRE system use double the number of lamps.)											Greatest Horse-Power on a 220 Volt Circuit.*	B. & S. Gauge Number.
		.45	.50	.55	.60	.65	.70	.75	.80	.90	1.00	1.10		
0000	303.	673	606	551	505	466	433	404	379	336	303	275	89.3	0000
000	254.	566	509	463	424	392	364	339	318	283	254	231	75.0	000
00	214.	476	428	389	357	329	306	285	267	238	214	195	63.1	00
0	180.	400	360	327	300	277	257	240	225	200	180	163	53.0	0
1	151.	336	302	275	252	232	216	201	189	168	151	137	44.5	1
2	127.	282	254	231	212	195	181	169	159	141	127	115	37.5	2
3	107.	237	213	194	178	164	152	142	133	119	107	97	31.4	3
4	90.	200	180	163	150	138	128	120	112	100	90	82	26.5	4
5	75.	167	151	137	125	116	107	100	94	84	75	68	22.2	5
6	63.	140	127	115	105	97	90	84	79	70	63	57	18.6	6
7	53.	118	106	97	89	82	76	71	66	59	53	48	15.7	7
8	45.	99	89	81	74	69	64	59	56	49	45	40	13.2	8
9	37.	83	75	68	62	57	53	50	47	41	37	34	11.0	9
10	31.	70	63	57	52	48	45	42	39	35	31	29	9.32	10
11	26.	59	53	48	44	41	38	35	33	29	26	24	7.81	11
12	22.	49	45	40	37	34	32	30	28	25	22	20	6.58	12
13	19.	42	38	34	31	29	27	25	23	21	19	17	5.54	13
14	16.	35	32	29	26	24	22	21	20	17	16	14	4.66	14
15	13.	29	26	24	22	20	19	17	16	14	13	12	3.89	15
16	11.	24	22	20	18	17	16	15	14	12	11	10	3.27	16
17	9.4	21	19	17	15	14	13	12	11	10	9	8	2.75	17
18	7.9	17	16	14	13	12	11	10	10	9	8	7	2.32	18
19	6.6	14	13	12	11	10	9	8	8	7	6	6	1.95	19
20	5.6	12	11	10	9	8	8	7	7	6	5	5	1.63	20
21	4.7	10	9	8	8	7	6	6	6	5	4	4	1.37	21
22	3.9	8	7	7	6	6	5	5	5	4	4	3	1.15	22

* These numbers represent ELECTRICAL HORSE-POWER; for MECHANICAL HORSE-POWER multiply these numbers by the efficiency of the motor.

TABLE OF HORSE-POWER EQUIVALENTS.

In wiring for motors, the wireman desires to know what current he must wire for, when the horse-power is given. To do this he must find the current corresponding to this horse-power. The horse-power tables as usually published are not well suited for this, as they are arranged for the reverse of this calculation. Furthermore, their ranges and the large number of decimals are far beyond the limits used by wiremen, and the tables are, therefore, unnecessarily large and cumbersome. The following table has therefore been prepared especially for wiremen, the ranges being chosen to cover those with which he has to deal, namely, from .1 to 30 H.P. and from 45 to 250 volts. It gives the currents in amperes required for different horse-powers at different voltages.

For horse-powers greater than the limit of the table, find the current for $\frac{1}{2}$, $\frac{1}{3}$, or $\frac{1}{4}$ of this horse-power, and then multiply the current obtained by 2, 3, or 4, respectively. For an odd number of horse-powers, as 21.5, for instance, add the current for 1.5 to that for 20 H.P.

For two, three, or four times the voltage given in the table, divide the current obtained from the table by two, three, or four, respectively.

The figures at the top may be read as amperes if those in the body of the table are read as volts. If many determinations are to be made for one particular voltage it is recommended to draw a red line on each side of that particular column.

For very large horse-powers, or when greater accuracy is required than is given in the table, the calculation should be performed. The current in amperes is equal to the horse-power multiplied by 746 and divided by the voltage.

These figures are for electrical horse-powers supplied to the motor. If the column of horse-powers is to represent mechanical horse-powers delivered by the motor, then divide the current obtained from the table by the efficiency of the motor (in units, thus, 70), and multiply by 100, which will give a proportionately greater current.

HORSE-POWER EQUIVALENTS IN VOLTS AND AMPERES.

Horse Power	VOLTS.															Horse Power
	45	50	55	60	65	70	75	80	85	90	95	100	105	110	115	
.1	1.66	1.49	1.36	1.24	1.15	1.07	.995	.932	.878	.829	.785	.746	.710	.678	.649	.1
.15	2.49	2.24	2.04	1.87	1.72	1.60	1.49	1.40	1.32	1.24	1.18	1.12	1.07	1.02	.973	.15
.2	3.32	2.99	2.71	2.49	2.30	2.13	1.99	1.87	1.76	1.65	1.57	1.49	1.42	1.36	1.30	.2
.25	4.15	3.73	3.39	3.11	2.87	2.67	2.49	2.33	2.20	2.07	1.96	1.87	1.78	1.70	1.62	.25
.3	4.97	4.48	4.07	3.73	3.44	3.20	2.99	2.80	2.63	2.49	2.36	2.24	2.13	2.04	1.95	.3
.35	5.80	5.22	4.75	4.35	4.02	3.73	3.48	3.26	3.07	2.90	2.75	2.61	2.49	2.37	2.27	.35
.4	6.63	5.97	5.43	4.97	4.59	4.26	3.98	3.73	3.51	3.32	3.14	2.98	2.84	2.71	2.60	.4
.45	7.46	6.71	6.10	5.59	5.16	4.80	4.48	4.20	3.95	3.73	3.53	3.36	3.20	3.05	2.92	.45
.5	8.29	7.46	6.78	6.22	5.74	5.33	4.97	4.66	4.39	4.15	3.93	3.73	3.55	3.39	3.24	.5
.55	9.12	8.21	7.46	6.84	6.31	5.86	5.47	5.13	4.83	4.56	4.32	4.10	3.91	3.73	3.57	.55
.6	9.95	8.95	8.14	7.46	6.89	6.40	5.97	5.59	5.27	4.97	4.71	4.48	4.26	4.07	3.89	.6
.65	10.8	9.70	8.82	8.08	7.46	6.93	6.46	6.06	5.71	5.39	5.10	4.85	4.62	4.41	4.22	.65
.7	11.6	10.5	9.49	8.70	8.03	7.46	6.96	6.53	6.14	5.80	5.50	5.22	4.97	4.75	4.54	.7
.75	12.4	11.2	10.2	9.32	8.61	7.99	7.46	6.99	6.58	6.22	5.89	5.59	5.33	5.09	4.86	.75
.8	13.3	11.9	10.9	9.95	9.18	8.52	7.96	7.46	7.02	6.63	6.28	5.97	5.68	5.42	5.19	.8
.85	14.1	12.7	11.5	10.6	9.76	9.06	8.45	7.93	7.46	7.05	6.68	6.34	6.04	5.76	5.51	.85
.9	14.9	13.4	12.2	11.2	10.3	9.59	8.95	8.39	7.90	7.46	7.07	6.71	6.39	6.10	5.84	.9
.95	15.8	14.2	12.9	11.8	10.9	10.1	9.45	8.86	8.34	7.87	7.46	7.09	6.75	6.44	6.16	.95
1.	16.6	14.9	13.6	12.4	11.5	10.7	9.95	9.32	8.78	8.29	7.85	7.46	7.10	6.78	6.49	1.
1.1	18.2	16.4	14.9	13.7	12.6	11.7	10.9	10.3	9.65	9.12	8.64	8.21	7.82	7.46	7.13	1.1
1.2	19.9	17.9	16.3	14.9	13.8	12.8	11.9	11.2	10.5	9.95	9.42	8.95	8.52	8.14	7.78	1.2
1.3	21.6	19.4	17.6	16.2	14.9	13.9	12.9	12.1	11.4	10.8	10.2	9.70	9.24	8.82	8.43	1.3
1.4	23.2	20.9	19.0	17.4	16.0	14.9	13.9	13.1	12.3	11.6	11.0	10.4	9.95	9.49	9.08	1.4
1.5	24.9	22.4	20.4	18.7	17.2	16.0	14.9	14.0	13.2	12.4	11.8	11.2	10.7	10.2	9.73	1.5
1.6	26.5	23.9	21.7	19.9	18.4	17.1	15.9	14.9	14.0	13.2	12.6	11.9	11.4	10.9	10.4	1.6
1.7	28.2	25.4	23.1	21.1	19.5	18.1	16.9	15.9	14.9	14.1	13.4	12.7	12.1	11.5	11.0	1.7
1.8	29.9	26.9	24.4	22.4	20.7	19.2	17.9	16.8	15.8	14.9	14.1	13.4	12.8	12.2	11.7	1.8
1.9	31.5	28.4	25.7	23.6	21.8	20.3	18.9	17.7	16.7	15.8	14.9	14.2	13.5	12.9	12.3	1.9
2.	33.2	29.9	27.1	24.9	23.0	21.3	19.9	18.7	17.6	16.5	15.7	14.9	14.2	13.6	13.0	2.
2.2	36.5	32.8	29.8	27.4	25.3	23.5	21.9	20.5	19.3	18.2	17.3	16.4	15.6	14.9	14.3	2.2
2.4	39.8	35.8	32.6	29.8	27.6	25.6	23.9	22.4	21.1	19.9	18.9	17.9	17.1	16.3	15.6	2.4
2.6	43.1	38.8	35.3	32.3	29.9	27.7	25.9	24.3	22.8	21.6	20.4	19.4	18.5	17.6	16.9	2.6
2.8	46.4	41.8	38.0	34.8	32.1	29.9	27.9	26.1	24.6	23.2	22.0	20.9	19.9	19.0	18.2	2.8
3.	49.7	44.8	40.7	37.3	34.4	32.0	29.9	28.0	26.3	24.9	23.6	22.4	21.3	20.4	19.5	3.
3.2	53.1	47.8	43.4	39.8	36.7	34.2	31.8	29.9	28.1	26.5	25.1	23.9	22.7	21.7	20.8	3.2
3.4	56.4	50.7	46.1	42.3	39.0	36.2	33.8	31.7	29.9	28.2	26.7	25.4	24.2	23.1	22.1	3.4
3.6	59.7	53.7	48.8	44.8	41.3	38.4	35.8	33.6	31.6	29.8	28.3	26.9	25.6	24.4	23.4	3.6
3.8	63.0	56.7	51.5	47.2	43.6	40.5	37.8	35.4	33.4	31.5	29.9	28.4	27.0	25.8	24.7	3.8
4.	66.3	59.7	54.3	49.7	45.9	42.6	39.8	37.3	35.1	33.2	31.4	29.8	28.4	27.1	26.0	4.
4.2	69.6	62.7	57.0	52.2	48.2	44.8	41.8	39.2	36.9	34.8	33.0	31.3	29.8	28.5	27.2	4.2
4.4	72.9	65.6	59.7	54.7	50.5	46.9	43.8	41.0	38.6	36.5	34.6	32.8	31.3	29.8	28.5	4.4
4.6	76.3	68.6	62.4	57.2	52.8	49.0	45.8	42.9	40.4	38.1	36.1	34.3	32.7	31.2	29.8	4.6
4.8	79.6	71.6	65.1	59.7	55.1	51.2	47.7	44.8	42.1	39.8	37.7	35.8	34.1	32.6	31.1	4.8
5.	82.9	74.6	67.8	62.2	57.4	53.3	49.7	45.6	43.9	41.5	39.3	37.3	35.5	33.9	32.4	5.
5.5	91.2	82.1	74.6	68.4	63.1	58.6	54.7	51.3	48.3	45.6	43.2	41.0	39.1	37.3	35.7	5.5
6.	99.5	89.5	81.4	74.6	68.9	64.0	59.7	55.9	52.7	49.7	47.1	44.8	42.6	40.7	38.9	6.
6.5	108.	97.0	88.2	80.8	74.6	69.3	64.6	60.6	57.1	53.9	51.0	48.5	46.2	44.1	42.2	6.5
7.	116.	105.	94.9	87.0	80.3	74.6	69.6	65.3	61.4	58.0	55.0	52.2	49.7	47.5	45.4	7.
7.5	124.	112.	102.	93.2	86.1	79.9	74.6	69.9	65.8	62.2	58.9	55.9	53.3	50.9	48.6	7.5
8.	133.	119.	109.	99.5	91.8	85.2	79.6	74.6	70.2	66.3	62.8	59.7	56.8	54.2	51.9	8.
8.5	141.	127.	115.	106.	97.6	90.6	84.5	79.3	74.6	70.5	66.8	63.4	60.4	57.6	55.1	8.5
9.	149.	134.	122.	112.	103.	95.9	89.5	83.9	79.0	74.6	70.7	67.1	63.9	61.0	58.4	9.
9.5	158.	142.	129.	118.	109.	101.	94.5	88.6	83.4	78.7	74.6	70.9	67.5	64.4	61.6	9.5
10.	166.	149.	136.	124.	115.	107.	99.5	93.2	87.8	82.9	78.5	74.6	71.0	67.8	64.9	10.
10.5	174.	157.	142.	131.	121.	112.	104.	97.9	92.1	87.1	82.4	78.4	74.6	71.2	68.1	10.5
11.	182.	164.	149.	137.	126.	117.	109.	103.	96.5	91.2	86.4	82.1	78.2	74.6	71.3	11.
11.5	191.	172.	156.	143.	132.	123.	114.	107.	101.	95.3	90.3	85.8	81.7	78.0	74.6	11.5
12.	199.	179.	163.	149.	138.	128.	119.	112.	105.	99.5	94.2	89.5	85.2	81.4	77.8	12.
12.5	207.	187.	170.	155.	144.	130.	124.	117.	110.	104.	98.2	93.3	88.8	84.8	81.1	12.5
13.	216.	194.	176.	162.	149.	139.	129.	121.	114.	108.	102.	97.0	92.4	88.2	84.3	13.
14.	232.	209.	190.	174.	160.	149.	139.	131.	123.	116.	110.	104.	99.5	94.9	90.8	14.
15.	249.	224.	204.	187.	172.	160.	149.	140.	132.	124.	118.	112.	107.	102.	97.3	15.
16.	265.	239.	217.	199.	184.	171.	159.	149.	140.	132.	126.	119.	114.	109.	104.	16.
17.	282.	254.	231.	211.	195.	181.	169.	159.	149.	141.	134.	127.	121.	115.	110.	17.
18.	299.	269.	244.	224.	207.	192.	179.	168.	158.	149.	141.	134.	128.	122.	117.	18.
19.	315.	284.	257.	236.	218.	203.	189.	177.	167.	158.	149.	142.	135.	129.	123.	19.
20.	332.	299.	271.	249.	230.	213.	199.	187.	176.	165.	157.	149.	142.	136.	130.	20.
22.	365.	328.	298.	274.	253.	235.	219.	205.	193.	182.	173.	164.	156.	149.	143.	22.
25.	415.	373.	339.	311.	287.	267.	249.	233.	220.	207.	196.	187.	178.	170.	162.	25.
30.	497.	448.	407.	373.	344.	320.	299.	280.	263.	249.	236.	224.	213.	204.	195.	30.
	45	50	55	60	65	70	75	80	85	90	95	100	105	110	115	

HORSE-POWER EQUIVALENTS IN VOLTS AND AMPERES.

Horse Power	120	130	140	150	160	170	180	190	200	210	220	230	240	250	Horse Power
						VOLTS.									
.1	.622	.574	.533	.497	.466	.439	.414	.393	.373	.355	.339	.324	.311	.298	.1
.15	.932	.861	.799	.746	.699	.658	.622	.589	.560	.533	.509	.487	.466	.448	.15
.2	1.24	1.15	1.07	.995	.932	.878	.829	.785	.746	.711	.678	.649	.622	.597	.2
.25	1.55	1.44	1.33	1.24	1.17	1.10	1.04	.982	.932	.888	.848	.811	.777	.746	.25
.3	1.87	1.72	1.60	1.49	1.40	1.32	1.24	1.18	1.12	1.07	1.02	.973	.932	.895	.3
.35	2.18	2.01	1.87	1.74	1.63	1.54	1.45	1.37	1.30	1.24	1.19	1.14	1.09	1.04	.35
.4	2.49	2.30	2.13	1.99	1.87	1.76	1.66	1.57	1.49	1.42	1.36	1.30	1.24	1.19	.4
.45	2.80	2.58	2.40	2.24	2.10	1.98	1.87	1.77	1.68	1.60	1.53	1.46	1.40	1.34	.45
.5	3.11	2.87	2.66	2.49	2.33	2.19	2.07	1.96	1.87	1.78	1.70	1.62	1.55	1.49	.5
.55	3.42	3.16	2.93	2.74	2.56	2.41	2.28	2.16	2.05	1.95	1.87	1.78	1.71	1.64	.55
.6	3.73	3.44	3.20	2.98	2.80	2.63	2.49	2.36	2.24	2.13	2.08	1.95	1.87	1.79	.6
.65	4.04	3.73	3.46	3.23	3.03	2.85	2.69	2.55	2.43	2.31	2.20	2.11	2.02	1.94	.65
.7	4.35	4.04	3.73	3.48	3.26	3.07	2.90	2.75	2.61	2.49	2.37	2.27	2.18	2.09	.7
.75	4.66	4.30	4.00	3.73	3.50	3.29	3.11	2.94	2.80	2.66	2.54	2.43	2.33	2.24	.75
.8	4.97	4.59	4.26	3.98	3.73	3.51	3.32	3.14	2.98	2.84	2.71	2.60	2.49	2.39	.8
.85	5.29	4.88	4.53	4.23	3.96	3.73	3.52	3.34	3.17	3.02	2.88	2.76	2.64	2.54	.85
.9	5.60	5.16	4.80	4.48	4.20	3.95	3.73	3.53	3.36	3.20	3.05	2.92	2.80	2.69	.9
.95	5.91	5.45	5.06	4.72	4.43	4.17	3.94	3.73	3.54	3.38	3.22	3.08	2.95	2.83	.95
1.	6.22	5.74	5.33	4.97	4.66	4.39	4.14	3.93	3.73	3.55	3.39	3.24	3.11	2.98	1.
1.1	6.84	6.31	5.86	5.47	5.13	4.83	4.56	4.32	4.10	3.91	3.73	3.57	3.42	3.28	1.1
1.2	7.46	6.89	6.39	5.97	5.60	5.27	4.97	4.71	4.48	4.26	4.07	3.89	3.73	3.58	1.2
1.3	8.08	7.46	6.93	6.46	6.06	5.71	5.39	5.10	4.85	4.62	4.41	4.22	4.04	3.88	1.3
1.4	8.70	8.03	7.46	6.96	6.53	6.14	5.80	5.50	5.22	4.97	4.75	4.54	4.35	4.18	1.4
1.5	9.32	8.61	7.99	7.46	6.99	6.58	6.22	5.89	5.60	5.33	5.09	4.87	4.66	4.48	1.5
1.6	9.95	9.18	8.52	7.96	7.46	7.02	6.63	6.28	5.97	5.68	5.43	5.19	4.97	4.77	1.6
1.7	10.6	9.75	9.06	8.45	7.92	7.46	7.05	6.68	6.34	6.04	5.77	5.51	5.28	5.07	1.7
1.8	11.2	10.3	9.59	8.95	8.39	7.90	7.46	7.07	6.71	6.40	6.11	5.84	5.60	5.37	1.8
1.9	11.8	10.9	10.1	9.45	8.86	8.34	7.87	7.46	7.09	6.75	6.44	6.16	5.91	5.67	1.9
2.	12.4	11.5	10.7	9.95	9.32	8.78	8.29	7.85	7.46	7.11	6.78	6.49	6.22	5.97	2.
2.2	13.7	12.6	11.7	10.9	10.3	9.65	9.12	8.64	8.20	7.82	7.46	7.14	6.84	6.56	2.2
2.4	14.9	13.8	12.8	11.9	11.2	10.5	9.95	9.42	8.95	8.52	8.14	7.78	7.46	7.16	2.4
2.6	16.2	14.9	13.9	12.9	12.1	11.4	10.8	10.2	9.70	9.24	8.82	8.43	8.08	7.76	2.6
2.8	17.4	16.1	14.9	13.9	13.1	12.3	11.6	11.0	10.4	9.95	9.49	9.08	8.70	8.36	2.8
3.	18.7	17.2	16.0	14.9	14.0	13.2	12.4	11.8	11.2	10.7	10.2	9.73	9.32	8.95	3.
3.2	19.9	18.4	17.1	15.9	14.9	14.0	13.3	12.6	11.9	11.4	10.9	10.4	9.95	9.55	3.2
3.4	21.1	19.5	18.1	16.9	15.9	14.9	14.1	13.4	12.7	12.1	11.5	11.0	10.6	10.1	3.4
3.6	22.4	20.7	19.2	17.9	16.8	15.8	14.9	14.1	13.4	12.8	12.2	11.7	11.2	10.7	3.6
3.8	23.6	21.8	20.2	18.9	17.7	16.7	15.8	14.9	14.2	13.5	12.9	12.3	11.8	11.3	3.8
4.	24.9	23.0	21.3	19.9	18.7	17.6	16.6	15.7	14.9	14.2	13.6	13.0	12.4	11.9	4.
4.2	26.1	24.1	22.4	20.9	19.6	18.4	17.4	16.5	15.7	14.9	14.2	13.6	13.1	12.5	4.2
4.4	27.4	25.2	23.5	21.9	20.5	19.3	18.2	17.3	16.4	15.6	14.9	14.3	13.7	13.1	4.4
4.6	28.6	26.4	24.5	22.9	21.5	20.2	19.1	18.1	17.2	16.3	15.6	14.9	14.3	13.7	4.6
4.8	29.9	27.6	25.6	23.9	22.4	21.1	19.9	18.8	17.9	17.1	16.3	15.6	14.9	14.3	4.8
5.	31.1	28.7	26.6	24.9	23.3	21.9	20.7	19.6	18.7	17.8	17.0	16.2	15.5	14.9	5.
5.5	34.2	31.6	29.3	27.4	25.6	24.1	22.8	21.6	20.5	19.5	18.7	17.8	17.1	16.4	5.5
6.	37.3	34.4	32.0	29.8	28.0	26.3	24.9	23.6	22.4	21.3	20.3	19.5	18.7	17.9	6.
6.5	40.4	37.3	34.6	32.3	30.3	28.5	26.9	25.5	24.3	23.1	22.0	21.1	20.2	19.4	6.5
7.	43.5	40.2	37.3	34.8	32.6	30.7	29.0	27.5	26.1	24.9	23.7	22.7	21.8	20.9	7.
7.5	46.6	43.0	40.0	37.3	35.0	32.9	31.1	29.4	28.0	26.6	25.4	24.3	23.3	22.4	7.5
8.	49.7	45.9	42.6	39.8	37.3	35.1	33.2	31.4	29.8	28.4	27.1	26.0	24.9	23.9	8.
8.5	52.9	48.8	45.3	42.3	39.6	37.3	35.2	33.4	31.7	30.2	28.8	27.6	26.4	25.4	8.5
9.	56.0	51.6	48.0	44.8	42.0	39.5	37.3	35.3	33.6	32.0	30.5	29.2	28.0	26.9	9.
9.5	59.1	54.5	50.6	47.2	44.3	41.7	39.4	37.3	35.4	33.8	32.2	30.8	29.5	28.3	9.5
10.	62.2	57.4	53.3	49.7	46.6	43.9	41.4	39.3	37.3	35.5	33.9	32.4	31.1	29.8	10.
10.5	65.3	60.3	56.0	52.2	49.0	46.1	43.5	41.2	39.2	37.3	35.6	34.1	32.6	31.3	10.5
11.	68.4	63.1	58.6	54.7	51.3	48.3	45.6	43.2	41.0	39.1	37.3	35.7	34.2	32.8	11.
11.5	71.5	66.0	61.3	57.2	53.6	50.5	47.7	45.2	42.9	40.9	39.0	37.3	35.7	34.3	11.5
12.	74.6	68.9	63.9	59.7	56.0	52.7	49.7	47.1	44.8	42.6	40.7	38.9	37.3	35.8	12.
12.5	77.7	71.7	65.6	62.2	58.3	54.9	51.8	49.1	46.6	44.4	42.4	40.5	38.9	37.3	12.5
13.	80.8	74.6	69.3	64.6	60.6	57.1	53.9	51.0	48.5	46.2	44.1	42.2	40.4	38.8	13.
14.	87.0	80.3	74.6	69.6	65.3	61.4	58.0	55.0	52.2	49.7	47.5	45.4	43.5	41.8	14.
15.	93.2	86.1	79.9	74.6	69.9	65.8	62.2	58.9	56.0	53.3	50.9	48.7	46.6	44.8	15.
16.	99.5	91.8	85.2	79.6	74.6	70.2	66.3	62.8	59.7	56.8	54.3	51.9	49.7	47.7	16.
17.	106.	97.5	90.6	84.5	79.2	74.6	70.5	66.8	63.4	60.4	57.7	55.1	52.8	50.7	17.
18.	112.	103.	95.9	89.5	83.9	79.0	74.6	70.7	67.1	64.0	61.1	58.4	55.9	53.7	18.
19.	118.	109.	101.	94.5	88.6	83.4	78.7	74.6	70.9	67.5	64.4	61.6	59.1	56.7	19.
20.	124.	115.	107.	99.5	93.2	87.8	82.9	78.6	74.6	71.1	67.8	64.9	62.2	59.7	20.
22.	137.	126.	117.	109.	103.	96.5	91.2	86.4	82.0	78.2	74.6	71.4	68.4	65.6	22.
25.	155.	144.	133.	124.	117.	110.	104.	98.2	93.2	88.8	84.8	81.1	77.7	74.6	25.
30.	187.	172.	160.	149.	140.	132.	124.	118.	112.	107.	102.	97.3	93.2	89.5	30.
	120	130	140	150	160	170	180	190	200	210	220	230	240	250	

WIRING TABLES.

The following set of five tables will be found very convenient for a special and limited class of work. They give the distances in feet up to 1,000, to which each size of wire of the B. & S. gauge will carry any given number of lamps at stated losses. Usually such tables are arranged differently, the sizes of wire being given for each number of lamp at regularly increasing distances. By the present arrangement, however, a table of the same size will cover a very much greater range of values; and, as it gives actual values instead of approximate ones, it is even more accurate, notwithstanding its increased range. It is also more convenient to use, because instead of following two rows of figures to their intersection, one line of figures is followed around a corner, which, for rapid work and a condensed table, is less confusing.

Such tables are necessarily limited to special lamps and losses. The values assumed in the following set have been chosen so as to cover as wide a range as possible, and to suit the usual lamps, voltages and losses. For lamps of slightly different currents than those assumed, it need be remembered merely, that if the current is slightly greater, the distances must be taken slightly less than those given, and *vice versa*. For half the losses given, take half the distances, or better, take the distances for double the number of lamps. Although calculated for five special cases, these tables may be used also for quite a number of other lamps, voltages and losses. These have all been classified in the index on the opposite page to facilitate finding which table to use.

It should be distinctly understood that these tables are not to be used for successive parts of branched circuits, unless the loss is understood to be for that part only. For instance, suppose the loss in a building is 2 per cent., and a certain circuit branches into two, say at one-fourth of the distance to the lamps, it is not correct to find the size of the first part for a two per cent. loss, and then the sizes of the second parts for a 2 per cent. loss, as this would give a total loss of 4 per cent. But if the loss on the first part be taken as, say $\frac{1}{2}$ per cent., and that on the second parts, the remaining $1\frac{1}{2}$ per cent., then the tables may be used for each part separately. This error has been made frequently by presumably reliable wiremen.

INDEX TO WIRING TABLES.

TWO WIRE SYSTEM.

For a	50 volt lamp, taking 1.1 amperes.	Loss 2.2	% or 1.1 volts, use table No. 1.
	50 " 1.1 "	" 4.4	2.2 " " 2.
	50 " 1.1 "	" 9.6	4.8 " " 3.
	50 " 1. "	" 2.	1. " " 1.
	50 " 1. "	" 4.	2. " " 2.
	50 " 1. "	" 8.8	4.4 " " 3.
For a	55 volt lamp, taking 1.1 amperes.	Loss 2.	% or 1.1 volts, use table No. 1.
	55 " 1.1 "	" 4.	2.2 " " 2.
	55 " 1.1 "	" 8.8	4.84 " " 3.
	55 " 1. "	" 1.8	1. " " 1.
	55 " 1. "	" 3.6	2. " " 2.
	55 " 1. "	" 8.	4.4 " " 3.
For a	75 volt lamp, taking .75 amperes.	Loss 1.	% or .75 volts, use table No. 1.
	75 " .75 "	" 2.	1.5 " " 2.
	75 " .75 "	" 4.4	3.3 " " 3.
	75 " .75 "	" 8.8	6.6 " " 4.
For a	75 volt lamp, taking .6 amperes.	Loss .8	% or .6 volts, use table No. 1.
	75 " .6 "	" 1.6	1.2 " " 2.
	75 " .6 "	" 3.5	2.64 " " 3.
	75 " .6 "	" 7.	5.3 (approx.) " 4.
For a	100 volt lamp, taking .5 amperes.	Loss .5	% or .5 volts, use table No. 1.
	100 " .5 "	" 1.	1. " " 2.
	100 " .5 "	" 2.2	2.2 " " 3.
	100 " .5 "	" 4.4	4.4 " " 4.
	100 " .5 "	" 8.8	8.8 " " 5.
For a	110 volt lamp, taking .5 amperes.	Loss .45	% or .5 volts, use table No. 1.
	110 " .5 "	" .9 (ap.) 1.	" " 2.
	110 " .5 "	" 2.	2.2 " " 3.
	110 " .5 "	" 4.	4.4 " " 4.
	110 " .5 "	" 8.	8.8 " " 5.
For a	110 volt lamp, taking .45 amperes.	Loss .41	% or .45 volts, use table No. 1.
	110 " .45 "	" .8 (ap.) .9	" " 2.
	110 " .45 "	" 1.8	2. (approx.) " 3.
	110 " .45 "	" 3.6	4. (approx.) " 4.
	110 " .45 "	" 7.2	8. (approx.) " 5.

THREE WIRE SYSTEM.

For a	100 volt lamp, taking .5 amperes.	Loss .55 % or .55 volts per lamp, use table No. 3.
	100 " .5 "	" 1.1	1.1 " " 4.
	100 " .5 "	" 2.2	2.2 " " 5.
For a	110 volt lamp, taking .5 amperes.	Loss .5 % or .55 volts per lamp, use table No. 3.
	110 " .5 "	" 1.	1.1 " " 4.
	110 " .5 "	" 2.	2.2 " " 5.
For a	110 volt lamp, taking .45 amperes.	Loss .45 % or .5 (approx.) use table No. 3.
	110 " .45 "	" .9	1. (approx.) " 4.
	110 " .45 "	" 1.8	2. (approx.) " 5.

MOTOR CURRENTS.

For a	50 volt circuit, and a loss of 2 % or 1. volt, use table No. 1.
	50 " " 4.	2. " 2.
	50 " " 8.8	4.4 " 3.
	50 " " 17.6	8.8 " 4.
For a	55 volt circuit, and a loss of 1.8 (ap.) 1. volt, use table No. 1.
	55 " " 3.6 (ap.) 2.	" 2.
	55 " " 8.	4.4 " 3.
	55 " " 16.	8.8 " 4.
For a	75 volt circuit, and a loss of 1.3 (ap.) 1. volt, use table No. 1.
	75 " " 2.7 (ap.) 2.	" 2.
	75 " " 5.9 (ap.) 4.4	" 3.
	75 " " 11.7 (ap.) 8.8	" 4.
For a	100 volt circuit, and a loss of 1. % or 1. volt, use table No. 1.
	100 " " 2.	2. " 2.
	100 " " 4.4	4.4 " 3.
	100 " " 8.8	8.8 " 4.
	100 " " 17.6	17.6 " 5.
For a	110 volt circuit, and a loss of .9 (ap.) 1. volt, use table No. 1.
	110 " " 1.8 (ap.) 2.	" 2.
	110 " " 4.	4.4 " 3.
	110 " " 8.	8.8 " 4.
	110 " " 16.	17.6 " 5.
For a	220 volt circuit, and a loss of .9 (ap.) 2. volt, use table No. 2.
	220 " " 2.	4.4 " 3.
	220 " " 4.	8.8 " 4.
	220 " " 8.	17.6 " 5.

WIRING TABLE No. 1.

Giving the maximum distances in feet up to 1000, to which each size of wire will carry any given number of lamps at the following losses:

For Motor Currents. (For "lamps" read "amperes.")

2 % loss for a 50 volt lamp taking 1 ampere.	Loss 1 volt.	2 % loss for a 50 volt circuit.
2 % loss for a 65 volt lamp taking 1.1 amperes.	Loss 1.1 volts.	1.3 % loss for a 75 volt circuit.
1.8 % loss for a 55 volt lamp taking 1. ampere.	Loss 1. volts.	1 % loss for a 100 volt circuit.
1 % loss for a 75 volt lamp taking .75 ampere.	Loss .75 volt.	.9 % loss for a 110 volt circuit.
.8 % loss for a 75 volt lamp taking .6 ampere.	Loss .6 volt.	
.5 % loss for a 100 volt lamp taking .5 ampere.	Loss .5 volt.	
.45 % loss for a 110 volt lamp taking .5 ampere.	Loss .5 volt.	
.41 % loss for a 110 volt lamp taking .45 ampere.	Loss .45 volt.	

DIRECTIONS: From the number of lamps at the top, follow downward to the required number of feet, thence to the right or left to the gauge number.

NUMBER OF LAMPS.

DISTANCES IN FEET.

HEATING LIMIT

For 1 ampere lamps is indicated thus: |
Small numbers are for .75 ampere lamps only.

B. & S. Gauge	1	2	3	4	5	6	7	8	9	10	12	14	16	18	20	22	25	30	35	40	45	50	55	60	65	70	80	90	100	B. & S. Gauge
16	122	61	41	30	24	20	17	15	14	12	10																			16
15	154	77	51	38	31	25	22	19	17	15	12	11																		15
14	194	97	65	48	39	32	28	24	22	19	16	14	12	11																14
13	244	122	81	61	49	41	35	30	27	24	20	17	15	14	12	11	10													13
12	308	154	103	77	62	51	44	38	34	31	26	22	19	17	15	14	12	10												12
11	388	194	129	97	78	65	56	48	43	39	32	28	24	22	19	18	16	13	11											11
10	490	245	163	122	98	82	70	61	54	49	41	35	30	27	25	22	20	16	14	12										10
9	617	309	206	164	123	103	88	77	69	62	51	44	38	34	31	28	25	21	18	16	14	12								9
8	778	389	269	195	156	130	111	97	87	78	65	56	48	43	39	35	31	26	22	20	17	16	14	13						8
7	982	491	327	245	196	164	140	123	109	98	82	70	61	54	49	45	39	33	28	25	22	20	18	16	15	14				7
6		619	413	309	248	206	177	155	138	124	103	89	77	69	62	56	50	41	35	31	28	25	23	21	19	18	16			6
5		780	520	390	312	260	223	195	174	156	130	111	97	87	77	71	62	52	45	39	35	31	28	26	24	22	20	17	16	5
4		984	657	492	394	328	281	246	219	197	164	140	123	109	98	89	70	66	49	44	39	36	33	31	28	25	22	20		4
3			827	620	496	414	355	310	276	248	206	177	155	138	124	113	99	83	71	62	55	50	45	41	39	36	31	28	25	3
2				783	626	522	447	391	348	313	261	224	196	175	157	142	125	104	89	78	70	63	57	52	48	45	39	35	31	2
1				987	790	658	564	494	439	395	329	281	246	219	197	179	158	132	113	99	88	79	72	66	61	56	49	44	40	1
0					995	829	711	622	553	497	415	355	310	277	249	226	199	166	143	124	111	100	91	83	77	71	62	55	50	0
00						996	784	697	626	523	446	382	349	314	285	261	206	179	157	140	126	114	105	97	90	78	70	63		00
000							989	880	781	656	565	495	440	396	360	317	264	226	196	176	158	144	132	122	113	99	88	79		000
0000								998	832	713	624	565	500	454	400	333	285	250	222	200	182	166	164	148	125	111	100			0000

Copyright, 1891, by CARL HERING.

THESE TABLES MUST NOT BE USED for successive parts of branched circuits, unless the loss represents only a PORTION of the total loss.

WIRING TABLE No. 2.

Giving the maximum distances in feet up to 1000, to which each size of wire will carry any given number of lamps at the following losses:

2 % loss for a 75 volt lamp taking .75 ampere.	Loss 1.5 volts.	
4 % loss for a 50 volt lamp taking 1. ampere.	Loss 2. volts.	
4 % loss for a 55 volt lamp taking 1.1 ampere.	Loss 2.2 volts.	
3.6 % loss for a 55 volt lamp taking 1. ampere.	Loss 2. volts.	
1.6 % loss for a 75 volt lamp taking .6 ampere.	Loss 1.2 volts.	
1 % loss for a 100 volt lamp taking .6 ampere.	Loss 1. volt.	
.9 % loss for a 110 volt lamp taking .5 ampere.	Loss 1. volt.	
.8 % loss for a 110 volt lamp taking .45 ampere.	Loss .9 volt.	

For Motor Currents. (For "lamps" read "amperes.")

4 % loss for a 50 volt circuit.
2.7 % loss for a 75 volt circuit.
2 % loss for a 100 volt circuit.
1.8 % loss for a 110 volt circuit.
.9 % loss for a 220 volt circuit.

DIRECTIONS: From the number of lamps at the top, follow downward to the required number of feet, thence to the right or left to the gauge number.

NUMBER OF LAMPS.

DISTANCES IN FEET.

HEATING: For 1 ampere lamps is indicated thus: | Small numbers are for .75 ampere lamps only.

B.&S. Gauge	1	2	3	4	5	6	7	8	9	10	12	14	16	18	20	22	25	30	35	40	45	50	55	60	65	70	80	90	100	B.&S. Gauge
16	244	122	81	61	49	41	34	30	27	24	20	17																		16
15	308	154	108	77	62	51	44	38	34	31	26	22	19																	15
14	388	194	129	97	78	65	56	48	43	39	32	28	24	22	19															14
13	486	244	163	122	98	81	70	61	54	49	41	35	30	27	24	22	20													13
12	616	308	205	164	123	103	88	77	69	62	51	44	38	34	31	28	25	21												12
11	776	388	258	194	155	129	112	97	87	78	65	55	48	43	39	35	31	26	22											11
10	980	490	327	245	196	163	140	122	109	98	82	70	61	54	49	45	39	33	28	25										10
9		617	413	306	247	206	176	154	138	123	103	88	77	69	62	57	50	41	36	31	28	26								9
8		778	519	389	311	259	222	195	174	156	130	111	97	87	78	71	62	52	45	39	36	31	28	26						8
7		982	656	491	393	327	281	245	219	196	164	140	123	109	98	89	79	66	56	49	44	39	36	33	31	29				7
6			827	619	496	413	355	309	276	248	206	177	155	124	113	99	88	71	62	55	50	45	41	39	38	31				6
5				780	624	520	447	390	349	312	260	228	195	174	166	141	124	104	89	78	70	62	57	52	48	46	39	35	31	5
4				984	788	657	564	492	439	394	328	281	246	219	197	178	158	132	113	98	88	79	72	66	61	56	49	44	39	4
3					992	827	711	620	553	496	414	355	310	276	248	225	165	141	125	110	99	91	83	77	71	62	55	50		3
2						896	783	697	626	522	447	391	348	318	283	250	208	178	156	139	125	114	105	97	90	78	70	63		2
1							987	880	790	658	564	494	439	395	359	316	263	226	197	175	157	144	131	121	112	99	88	79		1
0										996	829	711	622	553	497	452	398	331	283	248	220	198	181	164	153	142	124	110	100	0
00											896	784	697	627	570	502	418	359	314	279	251	228	209	193	178	157	140	126		00
000											969	880	791	720	633	526	452	396	352	317	288	264	244	226	198	176	158			000
0000												998	908	799	656	571	500	444	400	363	333	307	285	250	222	200				0000

THESE TABLES MUST NOT BE USED for successive parts of branched circuits, unless the loss represents only a PORTION of the total loss.

Copyright, 1891, by CARL HERING.

WIRING TABLE No. 3.

Giving the maximum distances in feet up to 1000, to which each size of wire will carry any given number of lamps at the following losses:

For the Two Wire System.

2 % loss for a 110 volt lamp taking .5 ampere.
8.8 % loss for a 50 volt lamp taking 1. ampere.
8.8 % loss for a 55 volt lamp taking 1. amperes.
8.8 % loss for a 55 volt lamp taking 1.1 amperes.
4.4 % loss for a 75 volt lamp taking 1. ampere.
4.4 % loss for a 75 volt lamp taking .75 ampere.
3.5 % loss for a 75 volt lamp taking .6 ampere.
2.2 % loss for a 100 volt lamp taking .5 ampere.
1.8 % loss for a 110 volt lamp taking .45 ampere.

For the Three Wire System.

.55 % loss for a 100 volt lamp taking .5 ampere. Loss .55 volt per lamp.
.5 % loss for a 110 volt lamp taking .5 ampere. Loss .55 volt per lamp.
.45 % loss for a 110 volt lamp taking .45 ampere. Loss .5 volt per lamp.

For Motor Currents. (For "lamps," read "amperes.")

8.8 % for a 50 volt circuit.   4 % for a 110 volt circuit.
5.9 % for a 75 volt circuit.   2 % for a 220 volt circuit.
4.4 % for a 100 volt circuit.

DIRECTIONS: From the number of lamps at the top, follow downward to the required number of feet, thence to the right or left to the gauge number.

For .75 ampere lamps is indicated thus: |
Small numbers are for .5 ampere lamps only.

NUMBER OF LAMPS — DISTANCES IN FEET.

B.&S. Gauge	1	2	3	4	5	6	7	8	9	10	12	14	16	18	20	22	25	30	35	40	45	50	55	60	65	70	80	90	100	B.&S. Gauge
16	536	268	179	134	107	89	76	67	59	54	44	38	33	30	27	26														16
15	675	338	225	169	135	112	96	84	75	68	56	48	42	38	34	31	27													15
14	852	426	284	213	170	142	121	106	95	85	71	60	53	48	43	39	34	28												14
13		538	368	269	215	179	153	134	119	108	89	77	67	60	54	49	43	36	31											13
12		677	451	338	271	225	198	169	150	136	112	96	84	75	68	62	54	46	39	34	30									12
11		854	569	427	342	284	244	213	190	171	142	122	106	95	85	78	68	57	49	43	38	34								11
10			718	538	431	359	307	269	239	215	179	153	134	120	108	98	86	72	62	54	48	43	40	38	36					10
9			906	680	544	453	388	340	302	272	226	194	170	151	136	123	109	91	78	68	60	54	50	46	42	39				9
8				856	685	571	489	428	380	343	285	244	214	190	171	156	137	114	98	86	76	69	63	57	53	49	43	38		8
7				864	720	617	540	480	432	360	309	270	240	216	195	173	144	123	108	99	86	79	72	66	62	60	48	43		7
6					908	778	681	605	546	454	389	340	302	272	248	218	182	156	136	121	109	99	91	84	78	68	66	55		6
5						981	858	763	687	572	491	429	361	343	312	275	229	196	172	153	137	126	114	105	98	86	76	69		5
4										862	722	619	541	481	433	394	346	389	247	210	192	173	167	144	133	123	108	96	87	4
3										910	780	683	607	546	496	437	364	312	273	243	218	198	182	156	136	121	109	96	109	3
2										968	860	765	686	626	550	459	393	344	306	275	260	229	211	196	172	153	138	174		2
1											965	868	790	695	579	496	434	386	347	315	289	267	246	221	193	183	174			1
0													905	876	730	625	547	486	439	398	365	337	313	274	243	219				0
00														910	788	690	613	552	501	460	425	394	345	306	276					00
000														995	870	774	695	632	580	535	497	435	386	348						000
0000															975	876	796	731	674	625	548	486	439							0000

HEAT LIMIT

THESE TABLES MUST NOT BE USED for successive parts of branched circuits, unless the loss represents only a PORTION of the total loss.

Copyright, 1891, by Carl Hering.

WIRING TABLE No. 4.

Giving the maximum distances in feet up to 1000, to which each size of wire will carry any given number of lamps at the following losses:

For the Three Wire System.
1 % loss for a 110 volt lamp taking .5 ampere.
1.1 % loss for a 100 volt lamp taking .5 ampere.
.9 % loss for a 110 volt lamp taking .45 ampere.

For the Two Wire System.
8.8 % loss for a 75 volt lamp taking .75 ampere. Loss 6.6 volts.
7. % loss for a 75 volt lamp taking .6 ampere. Loss 5.3 volts.
4.4 % loss for a 100 volt lamp taking .5 ampere. Loss 4.4 volts.
4. % loss for a 110 volt lamp taking .5 ampere. Loss 4.4 volts.
3.6 % loss for a 110 volt lamp taking .45 ampere. Loss 4. volts.

For Motor Currents.
17.6 % loss for a 50 volt circuit.
11.7 % loss for a 75 volt circuit.
8.8 % loss for a 100 volt circuit.
8.8 % loss for a 110 volt circuit.
4. % loss for a 220 volt circuit.

DIRECTIONS: From the number of lamps at the top, follow downward to the required number of feet, thence to the right or left to the gauge number.

NUMBER OF LAMPS — DISTANCES IN FEET.

B. & S. Gauge	2	3	4	5	6	7	8	9	10	12	14	16	18	20	25	30	35	40	45	50	55	60	65	70	75	80	85	90	100	B. & S. Gauge
16	536	368	268	215	179	153	134	119	107	89	76	67	59	54	43	36	31	27	24											16
15	675	450	338	270	225	193	169	150	135	112	96	84	75	68	54	45	39	34	30											15
14	852	568	426	341	284	243	213	190	170	142	121	106	95	86	68	57	49	43	38											14
13		716	538	430	358	307	269	239	216	179	153	134	119	108	86	72	61	54	48						See Note.					13
12		903	677	542	451	387	338	301	271	225	193	169	150	136	108	90	77	68	60		45	42	39	36	34	32	30			12
11			864	683	569	488	427	380	342	284	244	213	190	171	137	114	96	85	76	68	63	49	46	43	40	38	34			11
10			861	718	615	538	478	431	350	307	269	239	215	172	144	123	108	96	86	78	72	68	61	58	54	51	48	43		10
9			906	776	680	604	544	453	386	340	303	272	218	181	155	136	121	109	99	91	83	78	72	68	64	60				9
8				973	858	761	685	571	489	428	380	342	274	228	196	171	152	137	125	114	105	98	91	86	81	76	69			8
7					908	864	760	720	617	540	489	437	345	247	216	199	178	157	144	133	123	116	108	102	96	86				7
6						981	858	778	681	605	545	436	363	311	272	242	218	198	182	166	150	146	136	128	121	114	96			6
5									763	687	650	545	436	311	272	243	218	198	182	166	150	146	136	128	121	153	137			5
4					962	866	693	678	495	433	385	346	315	289	267	247	231	216	204	193	179								173	4
3						874	728	624	546	486	437	397	364	336	312	292	278	257	243	218									218	3
2						917	786	688	612	550	500	459	434	393	367	344	325	306	276										276	2
1						993	868	772	695	631	579	534	496	463	434	409	386	347											347	1
0							973	876	795	730	673	625	584	547	514	486	438												438	0
00								920	849	788	736	690	650	613	552														552	00
000									995	928	870	820	774	697															697	000
0000										975	877																		877	0000

NOTE:—HEATING LIMITS for .5 ampere lamps on the two wire system are indicated thus: |
The small figures to the right of these lines are for the three wire system only.

THESE TABLES MUST NOT BE USED for successive parts of branched circuits, unless the loss represents only a PORTION of the total loss.

WIRING TABLE No. 5.

Giving the maximum distances in feet up to 1000, to which each size of wire will carry any given number of lamps at the following losses:

For the Three Wire System.

2 % loss for a 110 volt lamp taking .5 ampere. Loss 2.2 volts per lamp.
2.2 % loss for a 100 volt lamp taking .5 ampere. Loss 2.2 volts per lamp.
1.8 % loss for a 110 volt lamp taking .45 ampere. Loss .4 volt per lamp.

For the Two Wire System.

8.8 % loss for a 100 volt lamp taking .5 ampere. Loss 8.8 volts.
8. % loss for a 110 volt lamp taking .5 ampere. Loss 8.8 volts.
7.2 % loss for a 110 volt lamp taking .45 ampere. Loss 7.9 volts.

For Motor Currents.

17.6 % loss for a 100 volt circuit.
16. % loss for a 110 volt circuit.
8. % loss for a 220 volt circuit.

DIRECTIONS: From the number of lamps at the top, follow downward to the required number of feet, thence to the right or left to the gauge number.

B. & S. Gauge.	NUMBER OF LAMPS.																												B. & S. Gauge.
	3	4	5	6	7	8	9	10	12	14	16	18	20	25	30	35	40	45	50	55	60	65	70	75	80	85	90	95	100	
	DISTANCES IN FEET.																													
16	715	536	429	358	306	268	238	215	179	153	134	119	107	86	72	61	54	48	43				H	E	A	T	I	N	G	16
15	901	675	540	450	386	338	300	270	225	193	169	150	135	108	90	77	68	60	54	50	45			L	I	M	I	T		15
14		852	682	568	487	426	379	341	284	243	213	190	170	136	114	97	85	76	68	62	57	53	49		See Note.					14
13			860	716	614	538	478	430	358	307	269	239	215	172	143	123	108	96	86	78	72	66	61	58	54	51				13
12				903	774	677	602	542	451	387	338	301	271	217	180	155	136	120	108	99	90	83	77	72	68	64	60	57	54	12
11					976	854	759	683	569	488	427	380	342	273	228	195	171	152	137	125	114	105	98	91	85	81	76	72	68	11
10							957	861	718	615	538	478	431	345	287	246	215	191	172	157	144	133	123	115	108	102	96	91	86	10
9									906	776	680	604	544	435	362	311	272	242	218	198	181	167	155	145	136	128	121	115	109	9
8	NOTE.—HEATING LIMITS for .5									978	856	761	685	548	456	391	342	304	274	249	228	211	196	183	171	161	152	144	137	8
7	ampere lamps on the three wire system are											960	864	691	576	494	432	384	345	314	288	266	247	230	216	203	192	182	173	7
6	indicated thus: ❘ The small figures to the right													871	726	622	545	484	436	396	363	335	311	291	272	256	242	229	218	6
5	of these lines are for .45 ampere lamps only.														916	785	687	610	550	500	458	423	392	366	343	324	305	289	275	5
4																990	866	770	693	630	578	533	495	462	433	408	385	365	346	4
3																		971	874	794	728	672	624	583	546	514	486	460	437	3
2																				999	917	847	786	734	688	648	612	580	550	2
1																							993	926	868	818	772	732	695	1
0																											972	922	875	0

THESE TABLES MUST NOT BE USED for successive parts of branched circuits, unless the loss represents only a **PORTION** of the total loss.

www.ingramcontent.com/pod-product-compliance
Lightning Source LLC
Chambersburg PA
CBHW022007190326
41519CB00010B/1421